# BBC DOCTOR WHO
# WHO GRAPHICA

**BBC**

# DOCTOR WHO

# WHO GR[A]PHICA

## AN INFOGRAPHIC GUIDE TO SPACE AND TIME

Simon Guerrier, Steve O'Brien and Ben Morris

HARPER DESIGN

An Imprint of HarperCollinsPublishers

This treeman represents the contents of WHOGRAPHICA. Its constituent parts are sized according to the lengths of the chapters.

# INTRODUCTION

*"I need more information."*
→ **The Twelfth Doctor,**
  **The Husbands of River Song (2015)**

In the *Doctor Who* story *Castrovalva* (1982), the newly regenerated Fifth Doctor recuperates in an alien town. But despite everyone being kind, he gets easily confused and talks nonsense – and he's sure there's something sinister going on.

At one point, he asks a nice doctor called Mergrave to draw a square. In this square he asks Mergrave to draw a map of the town. Then he asks, "Now, where's your pharmacy?"

Patiently, Mergrave marks it off on the map. "Up here, sir. And down here. And round here. And along here also."

Mergrave doesn't have four pharmacies – just one – but somehow it appears in four different places on the map. There's nothing wrong with the map: it's the town that is peculiar. But that wasn't obvious until Mergrave drew it.

THE DOCTOR'S COLOUR CODE

*Whographica* tries to do something similar: illustrating the universe of *Doctor Who* in ways that will draw out its richness, strangeness and wonder.

We've mined all 826 episodes from *An Unearthly Child* in 1963 to the 2015 Christmas special *The Husbands of River Song*, looking for the peculiar and unexpected and daft. In the process, we've rung up the Master to ask him how tall he is, discussed Daleks with the director of the Egypt Exploration Society and gone over and over 16 days, 8 hours, 13 minutes, 51 seconds' worth of episodes, looking for yet more boggling data.

In doing so, we hope to offer a captivating and intriguing journey through *Doctor Who* – for casual fans and hardened aficionados alike.

**SIMON, STEVE AND BEN**

Where possible, we've based the illustrations in this book on hard numbers. However, in some cases we've had to estimate or use our judgement – for example on exactly who counts as a companion.

# THE FIRST DOCTOR:
## VITAL STATISTICS

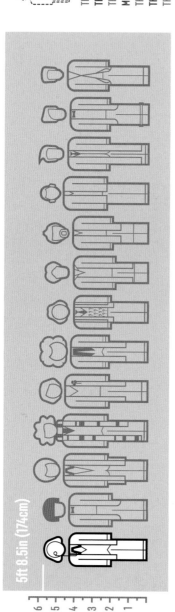

5ft 8.5in (174cm)

6
5
4
3
2
1

## EPISODES WHERE THE FIRST DOCTOR DOESN'T APPEAR

THE KEYS OF MARINUS: THE SCREAMING JUNGLE
THE KEYS OF MARINUS: THE SNOWS OF TERROR
THE DALEK INVASION OF EARTH: THE END OF TOMORROW
MISSION TO THE UNKNOWN
THE CELESTIAL TOYMAKER: THE HALL OF DOLLS
THE CELESTIAL TOYMAKER: THE DANCING FLOOR
THE TENTH PLANET EPISODE 3

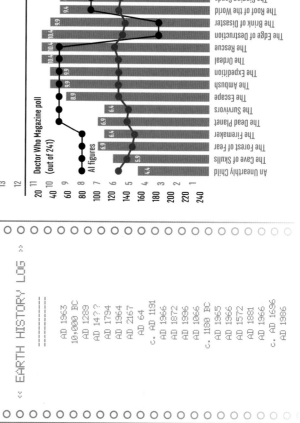

viewing figures (in millions)

Doctor Who Magazine poll (out of 241)

AI figures

100%
90%
80%
70%
60%
50%
40%
30%

20 / 40 / 10
60 / 9
80 / 8
100 / 7
120 / 6
140 / 5
160 / 4
180 / 3
200 / 2
220 / 1
240

An Unearthly Child
The Cave of Skulls
The Forest of Fear
The Firemaker
The Dead Planet
The Survivors
The Escape
The Ambush
The Expedition
The Ordeal
The Rescue
The Edge of Destruction
The Brink of Disaster
The Roof of the World
The Singing Sands
Five Hundred Eyes
The Wall of Lies
Rider from Shang-Tu
Mighty Kublai Khan
Assassin at Peking
The Sea of Death
The Velvet Web
The Screaming Jungle
The Snows of Terror
Sentence of Death
The Keys of Marinus
The Temple of Evil
The Warriors of Death
The Bride of Sacrifice
The Day of Darkness
Strangers in Space
The Unwilling Warriors
Hidden Danger
A Race Against Death
Kidnap
A Desperate Venture
A Land of Fear
Guests of Madame Guillotine
A Change of Identity
The Tyrant of France
A Bargain of Necessity
Prisoners of Conciergerie

## :: EARTH HISTORY LOG ::

AD 1963
10,000 BC
AD 1289
AD 14??
AD 1794
AD 1964
AD 2167
AD 64
c. AD 1191
AD 1966
AD 1872
AD 1996
AD 1066
c. 1180 BC
AD 1965
AD 1966
AD 1572
AD 1881
AD 1966
c. AD 1696
AD 1986

72%
28%

■ SCI-FI STORIES
■ PURELY HISTORICAL STORIES

*"Ah, yes! Thank you. It's good. Keep warm."*
— LAST WORDS

APPEARANCES IN OTHER STORIES

THE POWER OF THE DALEKS • DAY OF THE DALEKS • THE THREE DOCTORS
THE BRAIN OF MORBIUS • EARTHSHOCK
THE FIVE DOCTORS • RESURRECTION OF THE DALEKS
THE NEXT DOCTOR • THE ELEVENTH HOUR
THE VAMPIRES OF VENICE • VINCENT AND THE DOCTOR
THE LODGER • NIGHTMARE IN SILVER
THE NAME OF THE DOCTOR • THE DAY OF THE DOCTOR
THE WITCH'S FAMILIAR • THE ZYGON INVASION

*"What are you doing here?"*
— FIRST WORDS

# 134
NUMBER OF EPISODES

FASHION SHOW

Doctor Who Magazine poll (out of 241)

viewing figures (in millions)

| Story | Viewing figures |
|---|---|
| Checkmate | 8.3 |
| A Battle of Wits | 7.7 |
| The Meddling Monk | 8.8 |
| The Watcher | |
| The Planet of Decision | 9.5 |
| The Death of Doctor Who | 9 |
| Journey into Terror | 9.5 |
| Flight Through Eternity | 9 |
| The Death of Time | 9.5 |
| The Executioners | 10 |
| The Final Phase | 8.5 |
| The Search | 8.5 |
| The Dimensions of Time | 9.3 |
| The Space Museum | 10.5 |
| The Warlords | 9.5 |
| The Wheel of Fortune | |
| The Knight of Jaffa | 8.5 |
| The Lion | 10.5 |
| The Centre | 11.5 |
| Invasion | 12 |
| Crater of Needles | 13 |
| Escape to Danger | 12.5 |
| The Zarbi | 13 |
| The Web Planet | 13.5 |
| Inferno | 12 |
| Conspiracy | 10 |
| All Roads Lead to Rome | 11.5 |
| The Slave Traders | 13 |
| Desperate Measures | 13 |
| The Powerful Enemy | 12 |
| Flashpoint | 12.4 |
| The Waking Ally | 11.4 |
| The End of Tomorrow | 11.9 |
| Day of Reckoning | 11.9 |
| The Daleks | 12.4 |
| World's End | 11.4 |
| Crisis | 8.9 |
| Dangerous Journey | 8.4 |
| Planet of Giants | 8.4 |

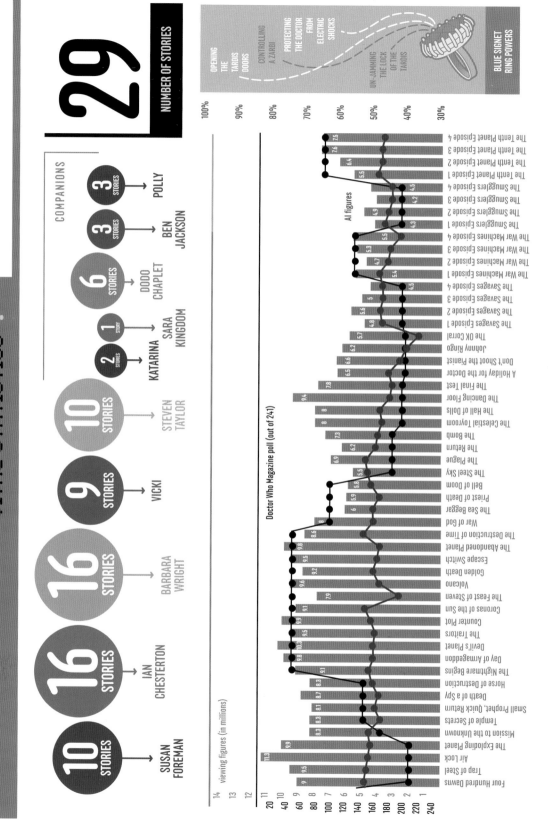

# THE FIRST DOCTOR
## VITAL STATISTICS

**NUMBER OF STORIES**

# 29

BLUE SIGNET RING POWERS

OPENING THE TARDIS DOORS

CONTROLLING A ZARBI

PROTECTING THE DOCTOR FROM ELECTRIC SHOCKS

UN-JAMMING THE LOCK OF THE TARDIS

COMPANIONS

**10 STORIES** — SUSAN FOREMAN

**16 STORIES** — IAN CHESTERTON

**16 STORIES** — BARBARA WRIGHT

**9 STORIES** — VICKI

**10 STORIES** — STEVEN TAYLOR

**2 STORIES** — KATARINA

**1 STORY** — SARA KINGDOM

**6 STORIES** — DODO CHAPLET

**3 STORIES** — BEN JACKSON

**3 STORIES** — POLLY

viewing figures (in millions)

Doctor Who Magazine poll (out of 241)

AI figures

Four Hundred Dawns
Trap of Steel
Air Lock
The Exploding Planet
Mission to the Unknown
Temple of Secrets
Small Prophet, Quick Return
Death of a Spy
Horse of Destruction
The Nightmare Begins
Day of Armageddon
Devil's Planet
The Traitors
Counter Plot
Coronas of the Sun
The Feast of Steven
Volcano
Golden Death
Escape Switch
The Abandoned Planet
The Destruction of Time
War of God
The Sea Beggar
Priest of Death
Bell of Doom
The Steel Sky
The Plague
The Return
The Bomb
The Celestial Toyroom
The Hall of Dolls
The Dancing Floor
The Final Test
A Holiday for the Doctor
Don't Shoot the Pianist
Johnny Ringo
The OK Corral
The Savages Episode 1
The Savages Episode 2
The Savages Episode 3
The Savages Episode 4
The War Machines Episode 1
The War Machines Episode 2
The War Machines Episode 3
The War Machines Episode 4
The Smugglers Episode 1
The Smugglers Episode 2
The Smugglers Episode 3
The Smugglers Episode 4
The Tenth Planet Episode 1
The Tenth Planet Episode 2
The Tenth Planet Episode 3
The Tenth Planet Episode 4

THE FIRST
DOCTOR

THE TARDIS DEMATERIALISING
(FROM OUTSIDE)

A DALEK

CYBERMEN

THE SECOND DOCTOR AND
CHANGE OF APPEARANCE

THE ICE
WARRIORS

LETHBRIDGE-
STEWART

THE SONIC
SCREWDRIVER

USE OF THE WORD
'TIME LORD'

THE THIRD
DOCTOR

SILURIANS

THE MASTER

A SONTARAN

THE FOURTH
DOCTOR

DAVROS

ZYGONS

K-9

THE FIFTH
DOCTOR

THE MARA

THE SIXTH
DOCTOR

THE RANI

THE SEVENTH
DOCTOR

THE SPECIAL
WEAPONS DALEK

THE EIGHTH
DOCTOR

THE NINTH
DOCTOR

PSYCHIC
PAPER

THE TENTH
DOCTOR

THE WEEPING
ANGELS

RIVER
SONG

THE ELEVENTH
DOCTOR

THE WAR
DOCTOR

THE SILENCE

THE TWELFTH
DOCTOR

SONIC
GLASSES

# FIRSTS

When did we originally clap eyes on...?

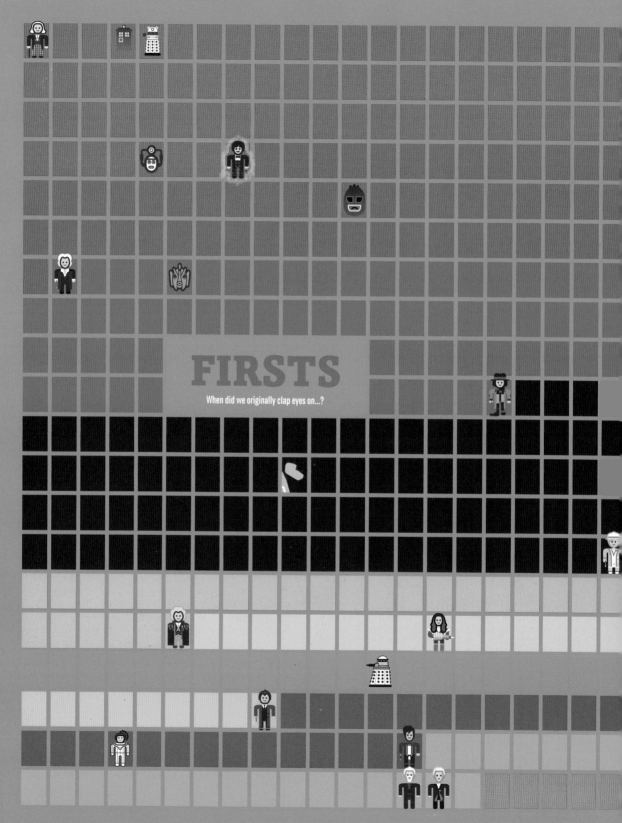

# FIRSTS

When did we originally clap eyes on...?

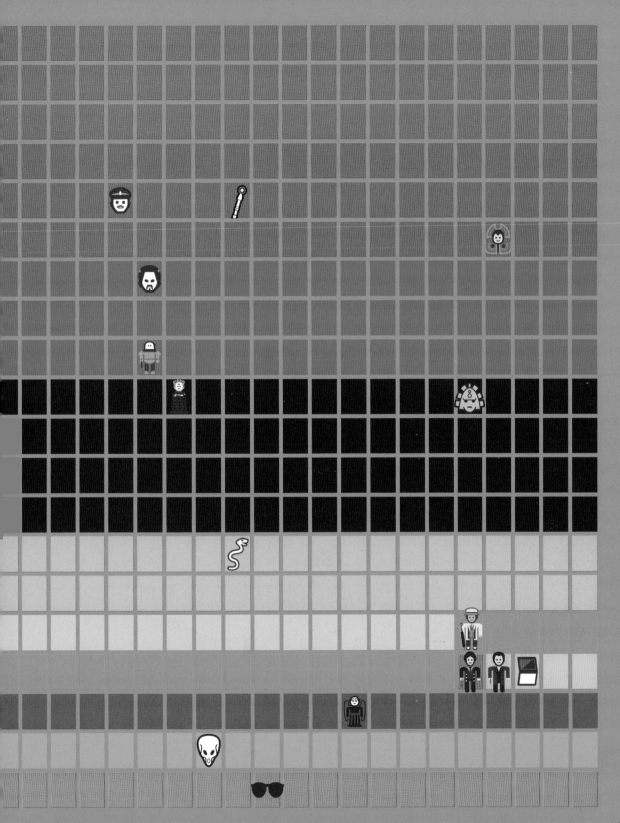

# A Division of Doctors

Their costume choices and the monsters we've seen them fight.

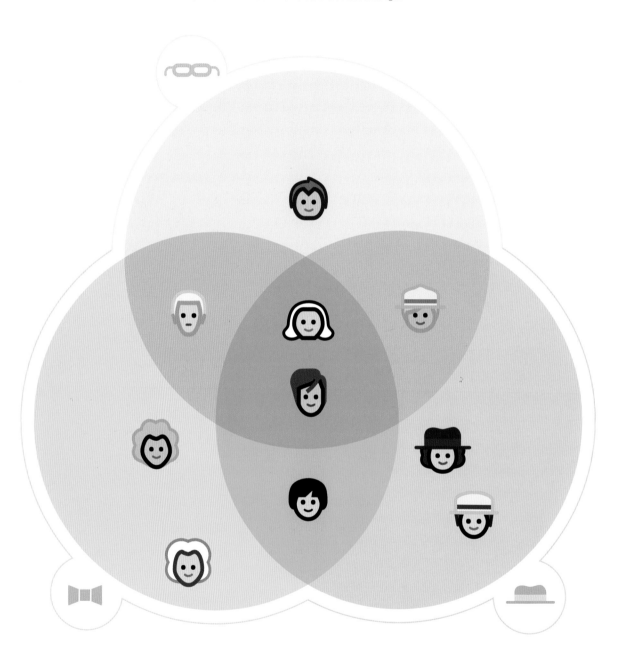

Sartorial choices of the Doctor's different incarnations.

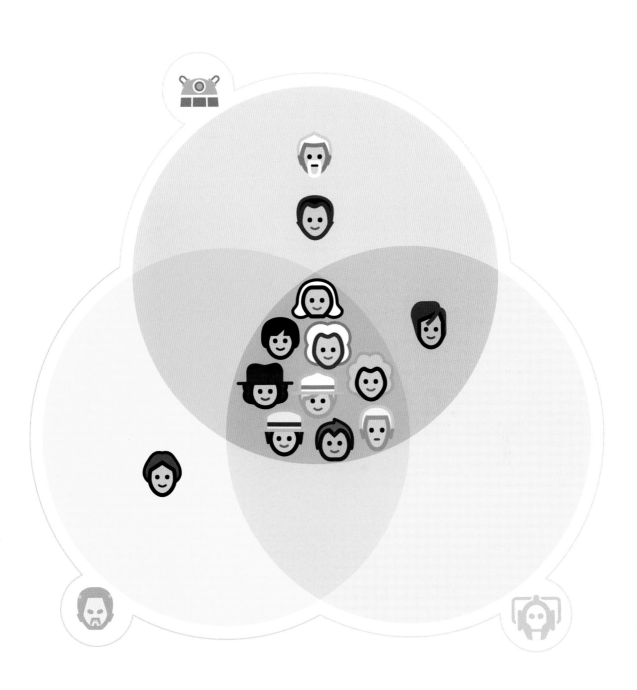

Villains fought by the Doctor's different incarnations.

# The Doctor's Age

"How far, Doctor? How long have you lived?"
Well, evil Time Lord Morbius, that's a tricky question…

## 01

The First Doctor was 8 when he entered the Academy. The Sound of Drums
He was a 'kid of 90' when he first visited the Medusa Castle. The Stolen Earth
The Doctor stole the TARDIS when he was around the age of 200.
The Impossible Astronaut, The Doctor's Wife

## 02

The Second Doctor tells Victoria that he is 'around' 450.
The Tomb of the Cybermen

## 03

The Third Doctor suggests he's several thousand years old.
Doctor Who and The Silurians, The Mind of Evil

## 04

The Fourth Doctor is around 750. Pyramids of Mars, The Brain of Morbius
He claims to be 756, but Romana shops his real age: 759. The Ribos Operation

## 05

The Fifth Doctor makes no reference to his age.

## 06

The Sixth Doctor is 900 years old when travelling with Peri.
Revelation of the Daleks

## 07

The Seventh Doctor is 953, the same age as the Rani.
Time and the Rani

## 08

The Eighth Doctor makes no reference to his age.

## WAR

The War Doctor is 400 years younger than the Eleventh Doctor, making him between 800 and 900 years old.
The Day of the Doctor

## 09

The Ninth Doctor is 900.
Aliens of London

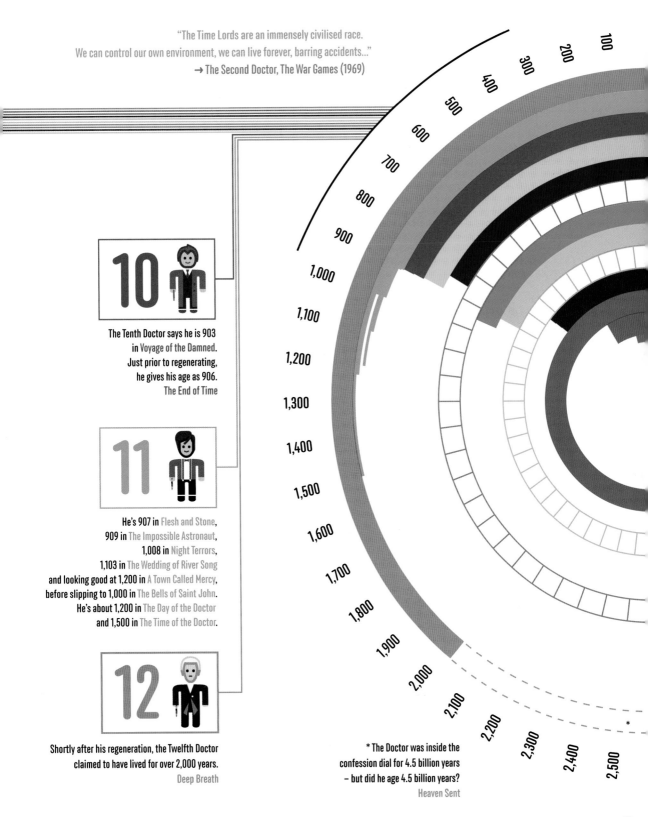

"The Time Lords are an immensely civilised race.
We can control our own environment, we can live forever, barring accidents..."
→ The Second Doctor, The War Games (1969)

100
200
300
400
500
600
700
800
900
1,000
1,100
1,200
1,300
1,400
1,500
1,600
1,700
1,800
1,900
2,000
2,100
2,200
2,300
2,400
2,500

**10**

The Tenth Doctor says he is 903
in Voyage of the Damned.
Just prior to regenerating,
he gives his age as 906.
The End of Time

**11**

He's 907 in Flesh and Stone,
909 in The Impossible Astronaut,
1,008 in Night Terrors,
1,103 in The Wedding of River Song
and looking good at 1,200 in A Town Called Mercy,
before slipping to 1,000 in The Bells of Saint John.
He's about 1,200 in The Day of the Doctor
and 1,500 in The Time of the Doctor.

**12**

Shortly after his regeneration, the Twelfth Doctor
claimed to have lived for over 2,000 years.
Deep Breath

\* The Doctor was inside the
confession dial for 4.5 billion years
– but did he age 4.5 billion years?
Heaven Sent

# Return of the Doctor

How long it took former incarnations to appear on screen again.*

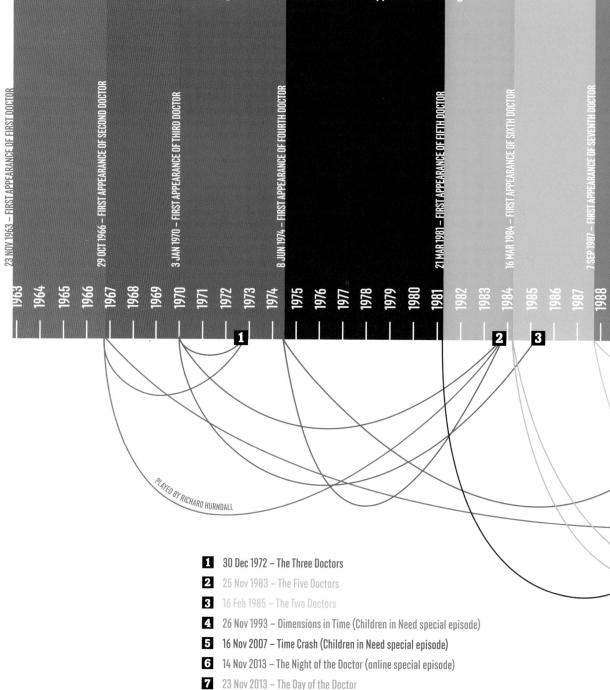

23 NOV 1963 – FIRST APPEARANCE OF FIRST DOCTOR
29 OCT 1966 – FIRST APPEARANCE OF SECOND DOCTOR
3 JAN 1970 – FIRST APPEARANCE OF THIRD DOCTOR
8 JUN 1974 – FIRST APPEARANCE OF FOURTH DOCTOR
21 MAR 1981 – FIRST APPEARANCE OF FIFTH DOCTOR
16 MAR 1984 – FIRST APPEARANCE OF SIXTH DOCTOR
7 SEP 1987 – FIRST APPEARANCE OF SEVENTH DOCTOR

1963 1964 1965 1966 1967 1968 1969 1970 1971 1972 1973 1974 1975 1976 1977 1978 1979 1980 1981 1982 1983 1984 1985 1986 1987 1988

PLAYED BY RICHARD HURNDALL

**1** 30 Dec 1972 – The Three Doctors
**2** 25 Nov 1983 – The Five Doctors
**3** 16 Feb 1985 – The Two Doctors
**4** 26 Nov 1993 – Dimensions in Time (Children in Need special episode)
**5** 16 Nov 2007 – Time Crash (Children in Need special episode)
**6** 14 Nov 2013 – The Night of the Doctor (online special episode)
**7** 23 Nov 2013 – The Day of the Doctor
**8** 23 Aug 2014 – Deep Breath

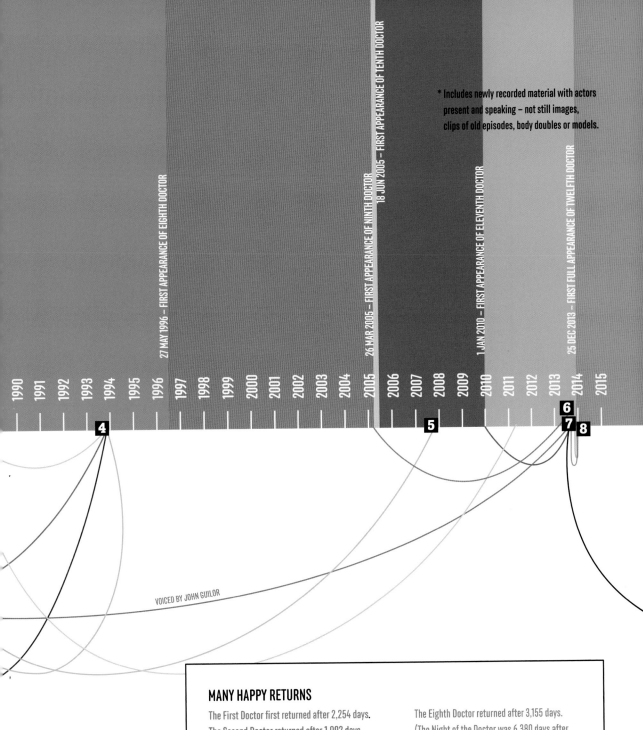

* Includes newly recorded material with actors present and speaking – not still images, clips of old episodes, body doubles or models.

27 MAY 1996 – FIRST APPEARANCE OF EIGHTH DOCTOR

26 MAR 2005 – FIRST APPEARANCE OF NINTH DOCTOR

18 JUN 2005 – FIRST APPEARANCE OF TENTH DOCTOR

1 JAN 2010 – FIRST APPEARANCE OF ELEVENTH DOCTOR

25 DEC 2013 – FIRST FULL APPEARANCE OF TWELFTH DOCTOR

1990 1991 1992 1993 1994 1995 1996 1997 1998 1999 2000 2001 2002 2003 2004 2005 2006 2007 2008 2009 2010 2011 2012 2013 2014 2015

**4** **5** **6** **7** **8**

VOICED BY JOHN GUILOR

## MANY HAPPY RETURNS

The First Doctor first returned after 2,254 days.
The Second Doctor returned after 1,092 days.
The Third Doctor returned after 3,457 days.
**The Fourth Doctor returned after 4,633 days.**
The Fifth Doctor returned after 3,542 days.
The Sixth Doctor returned after 2,272 days.
The Seventh Doctor has yet to return.

The Eighth Doctor returned after 3,155 days.
(The Night of the Doctor was 6,380 days after
his last appearance as the "current" Doctor.)
The Ninth Doctor has yet to return.
**The Tenth Doctor returned after 1,422 days.**
The Eleventh Doctor returned after 241 days.
The Twelfth Doctor returned after -32 days.

# Time and Relatives

The happy, sprawling family of Doctor Who...

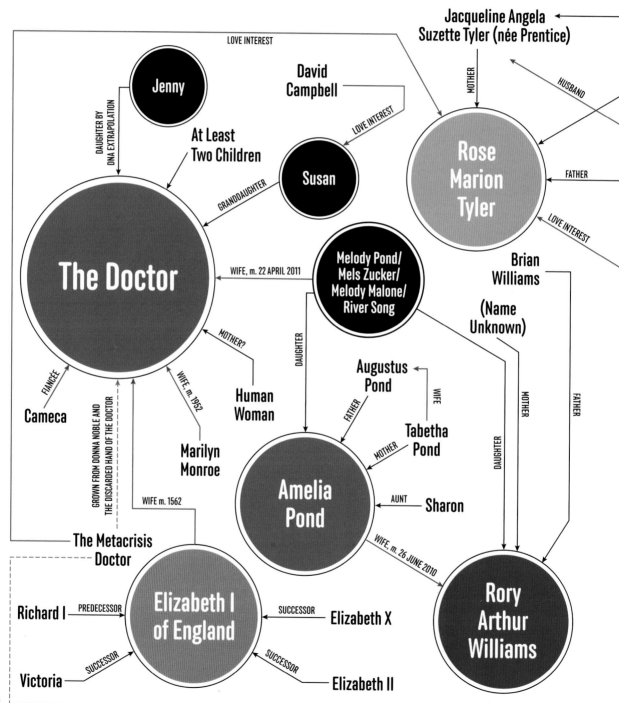

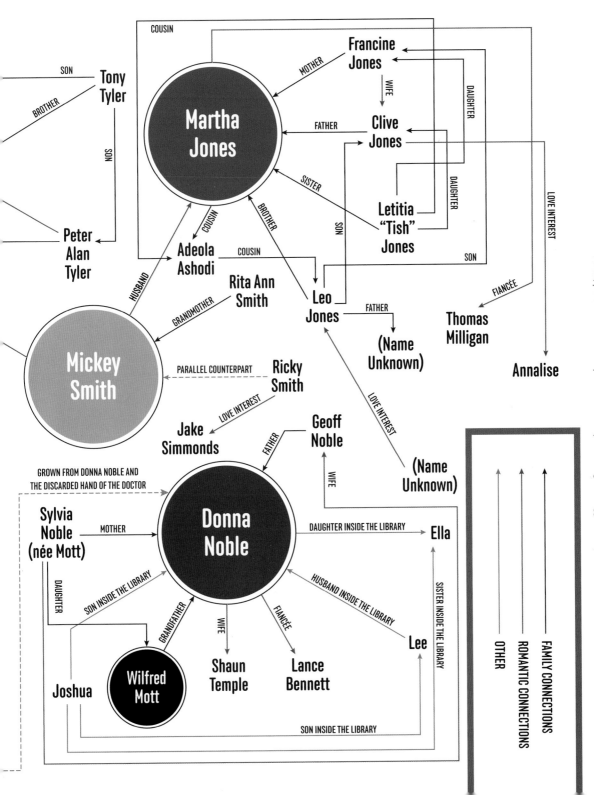

# Oh, I Dig Your FAB Gear!

Where the Doctor got his clothes...

The Eleventh Doctor's glasses were given to him by Amy Pond.

The Second Doctor's clothes (in The Power of the Daleks) regenerated with him!

The Tenth Doctor's coat was given to him by Janis Joplin.

The Eleventh Doctor's fez in The Day of the Doctor was stolen from the National Gallery.

The Eleventh Doctor's Stetson was given to him by Craig Owens.

The Eighth Doctor's original shoes were given to him by Dr Grace Holloway (they belonged to her ex-boyfriend, Brian).

The Third Doctor's clothes were inspired by those stolen from a hospital in Ashbridge.

"I'm saving the world: I need a decent shirt. To hell with the raggedy. Time to put on a show."
→ The Eleventh Doctor, The Eleventh Hour (2010)

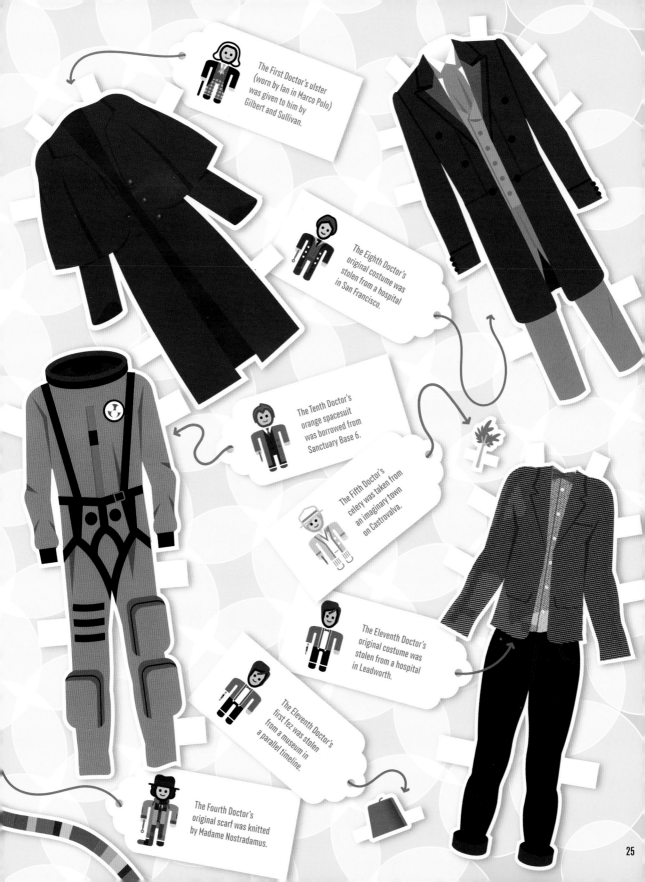

The First Doctor's ulster (worn by Ian in Marco Polo) was given to him by Gilbert and Sullivan.

The Eighth Doctor's original costume was stolen from a hospital in San Francisco.

The Tenth Doctor's orange spacesuit was borrowed from Sanctuary Base 6.

The Fifth Doctor's celery was taken from an imaginary town on Castrovalva.

The Eleventh Doctor's original costume was stolen from a hospital in Leadworth.

The Eleventh Doctor's first fez was stolen from a museum in a parallel timeline.

The Fourth Doctor's original scarf was knitted by Madame Nostradamus.

# THE DOCTOR'S
# ANATOMY

He might look like a human, but there's
more to the Doctor than meets the eye...

The Third Doctor has
a tattoo of a snake.

Some psychic ability:
can sense danger,
evil and/or Daleks.
Also sensitive to jumps
in time and how the
future will play out.

Able to regenerate all his cells at once,
transforming into a new person.
Usually, Time Lords can do this 12 times,
but the Doctor has been given a
whole new regeneration cycle.

Amazing memory, knows 5 billion
languages including baby and horse.
Brain more complex than a human's.
Left and right sides work in unison
via a specialised neural ganglia,
combining data storage and retrieval
with logical interest and the intuitive leap.
Until he left Time Lord society, a reflex
link let him tune into the thousand
super-brains of the Time Lord intelligentsia.

Platelet stickiness does not
match any human blood group.

Non-human metabolism:
an aspirin would probably kill him,
but he can also detoxify himself
using protein, salt and a shock
to stimulate inhibited
enzymes into reversal.

Clothes and shoes can
change during regeneration.

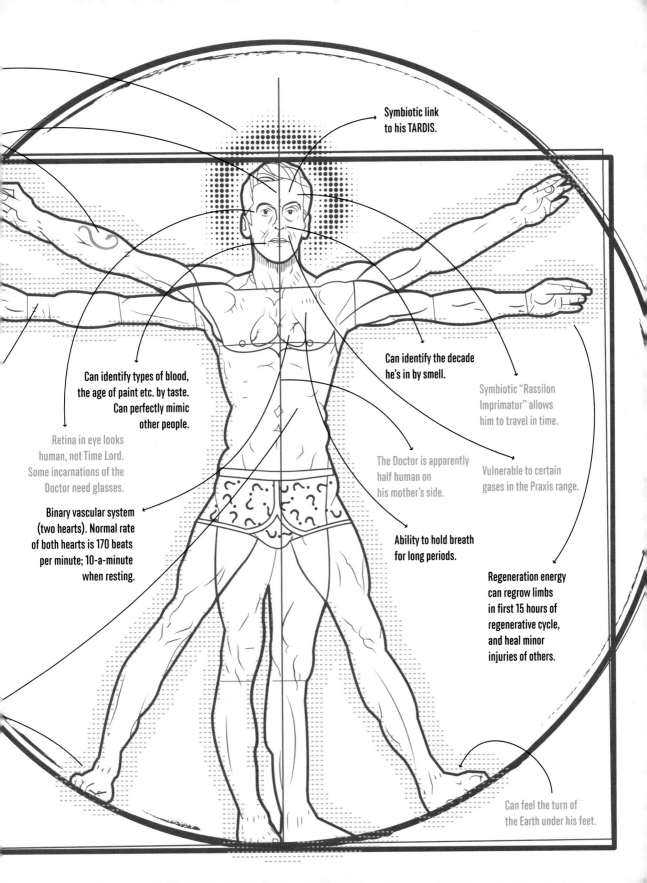

Symbiotic link to his TARDIS.

Can identify types of blood, the age of paint etc. by taste. Can perfectly mimic other people.

Retina in eye looks human, not Time Lord. Some incarnations of the Doctor need glasses.

Binary vascular system (two hearts). Normal rate of both hearts is 170 beats per minute; 10-a-minute when resting.

Can identify the decade he's in by smell.

Symbiotic "Rassilon Imprimatur" allows him to travel in time.

The Doctor is apparently half human on his mother's side.

Vulnerable to certain gases in the Praxis range.

Ability to hold breath for long periods.

Regeneration energy can regrow limbs in first 15 hours of regenerative cycle, and heal minor injuries of others.

Can feel the turn of the Earth under his feet.

# THE SECOND DOCTOR
## VITAL STATISTICS

Legend:
- 14 viewing figures (in millions)
- 13 Doctor Who Magazine poll (out of 241)
- 12 AI figures

Chart (AI figures):

| Episode | Value |
|---|---|
| The Power of the Daleks Episode 1 | 7.9 |
| The Power of the Daleks Episode 2 | 7.8 |
| The Power of the Daleks Episode 3 | 7.5 |
| The Power of the Daleks Episode 4 | 7.8 |
| The Power of the Daleks Episode 5 | 8 |
| The Power of the Daleks Episode 6 | 7.8 |
| The Highlanders Episode 1 | 6.7 |
| The Highlanders Episode 2 | 6.8 |
| The Highlanders Episode 3 | 7.4 |
| The Highlanders Episode 4 | 7.3 |
| The Underwater Menace Episode 1 | 8.3 |
| The Underwater Menace Episode 2 | 7.5 |
| The Underwater Menace Episode 3 | 7.1 |
| The Underwater Menace Episode 4 | 7 |
| The Moonbase Episode 1 | 8.1 |
| The Moonbase Episode 2 | 8.9 |
| The Moonbase Episode 3 | 8.2 |
| The Moonbase Episode 4 | 8.1 |
| The Macra Terror Episode 1 | 8 |
| The Macra Terror Episode 2 | 7.9 |
| The Macra Terror Episode 3 | 8.5 |
| The Macra Terror Episode 4 | 8.4 |
| The Faceless Ones Episode 1 | 8 |
| The Faceless Ones Episode 2 | 6.4 |
| The Faceless Ones Episode 3 | 7.9 |
| The Faceless Ones Episode 4 | 6.9 |
| The Faceless Ones Episode 5 | 7.1 |
| The Faceless Ones Episode 6 | 8 |
| The Evil of the Daleks Episode 1 | 4.1 |
| The Evil of the Daleks Episode 2 | 7.5 |
| The Evil of the Daleks Episode 3 | 6.1 |
| The Evil of the Daleks Episode 4 | 5.3 |
| The Evil of the Daleks Episode 5 | 5.1 |
| The Evil of the Daleks Episode 6 | 6.8 |
| The Evil of the Daleks Episode 7 | 6.1 |

Axis (viewing figures): 20, 40, 60, 80, 100, 120, 140, 160, 180, 200, 220, 240

Axis (percentage): 30%, 40%, 50%, 60%, 70%, 80%, 90%, 100%

## MONSTER ROLL-CALL

- DALEKS: 2 STORIES
- CYBERMEN: 4 STORIES
- YETI: 2 STORIES
- ICE WARRIORS: 2 STORIES

## REJECTED COSTUMES

## NUMBER OF STORIES THAT ENDED ON A CLIFFHANGER

6

## SONIC SCREWDRIVER MARK I

"Slower. Slower!
Concentrate on one thing."
### FIRST WORDS

**HATS**

**FASHION SHOW**

**Noms de Plume**

Doktor von Wer
(The Highlanders)

Dr W
(The Underwater Menace)

John Smith
(Wheel in Space)
(The War Games)

---

**MENTIONS OF "WHEN I SAY RUN, RUN!"**

THE POWER OF THE DALEKS
THE FACELESS ONES
THE EVIL OF THE DALEKS
THE TOMB OF THE CYBERMEN
THE ABOMINABLE SNOWMEN

**MENTIONS OF "OH MY WORD!"**

THE ICE WARRIORS
THE DOMINATORS
THE MIND ROBBER
THE INVASION
THE SEEDS OF DEATH
THE SPACE PIRATES
THE WAR GAMES
THE THREE DOCTORS
THE TWO DOCTORS

*"No! Stop!
You're making me giddy.
No, you can't do this to me.
No, No, no, no..."*

**LAST WORDS**

---

14
13
12    viewing figures (in millions)
    Doctor Who Magazine poll (out of 241)

All figures

100%
90%
80%
70%
60%
50%
40%
30%

20 11
40 10
60 9
80 8
100 7
120 6
140 5
160 4
180 3
200 2
220 1
240

The Tomb of the Cybermen Episode 1 — 6
The Tomb of the Cybermen Episode 2 — 6.4
The Tomb of the Cybermen Episode 3 — 7.2
The Tomb of the Cybermen Episode 4 — 7.4
The Abominable Snowmen Episode 1 — 6.3
The Abominable Snowmen Episode 2 — 6
The Abominable Snowmen Episode 3 — 7.1
The Abominable Snowmen Episode 4 — 7.1
The Abominable Snowmen Episode 5 — 7.2
The Abominable Snowmen Episode 6 — 7.4
The Ice Warriors Episode 1 — 6.7
The Ice Warriors Episode 2 — 7.1
The Ice Warriors Episode 3 — 7.4 7.3
The Ice Warriors Episode 4 — 7.3
The Ice Warriors Episode 5 — 8
The Ice Warriors Episode 6 — 7.5
The Enemy of the World Episode 1 — 6.8
The Enemy of the World Episode 2 — 7.6
The Enemy of the World Episode 3 — 7.1
The Enemy of the World Episode 4 — 7.8
The Enemy of the World Episode 5 — 6.9
The Enemy of the World Episode 6 — 8.3
The Web of Fear Episode 1 — 7.2
The Web of Fear Episode 2 — 6.8
The Web of Fear Episode 3 — 7
The Web of Fear Episode 4 — 8.4
The Web of Fear Episode 5 — 8
The Web of Fear Episode 6 — 8.3
Fury from the Deep Episode 1 — 8.2
Fury from the Deep Episode 2 — 7.9
Fury from the Deep Episode 3 — 7.7
Fury from the Deep Episode 4 — 6.6
Fury from the Deep Episode 5 — 5.9
Fury from the Deep Episode 6 — 6.9
The Wheel in Space Episode 1 — 6.9 7.2
The Wheel in Space Episode 2 — 6.9
The Wheel in Space Episode 3 — 7.5
The Wheel in Space Episode 4 — 8.6
The Wheel in Space Episode 5 — 6.8
The Wheel in Space Episode 6 — 6.5

31

# THE SECOND DOCTOR

## VITAL STATISTICS

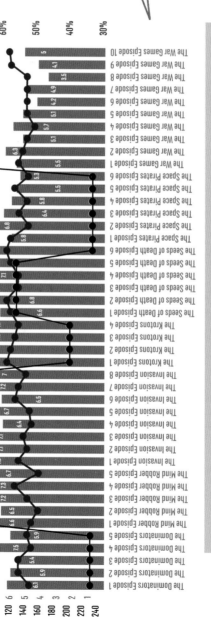

THE MACRA TERROR
→ THE DOCTOR'S FACE APPEARS FOR THE FIRST TIME IN THE TITLE SEQUENCE

MENTIONS OF "OH MY GIDDY AUNT!"

THE KROTONS
THE THREE DOCTORS
THE TWO DOCTORS

# 119

NUMBER OF EPISODES

5ft 9in (175cm)

14  viewing figures (in millions)
13
12  Doctor Who Magazine poll (out of 241)

AI figures

| Episode | viewing figures |
|---|---|
| The Dominators Episode 1 | 6.1 |
| The Dominators Episode 2 | 5.9 |
| The Dominators Episode 3 | 5.4 |
| The Dominators Episode 4 | 5.9 |
| The Dominators Episode 5 | 7.5 |
| The Mind Robber Episode 1 | 6.6 |
| The Mind Robber Episode 2 | 6.5 |
| The Mind Robber Episode 3 | 7.2 |
| The Mind Robber Episode 4 | 7.3 |
| The Mind Robber Episode 5 | 6.7 |
| The Invasion Episode 1 | 7.3 |
| The Invasion Episode 2 | 7.1 |
| The Invasion Episode 3 | 7.1 |
| The Invasion Episode 4 | 6.4 |
| The Invasion Episode 5 | 6.7 |
| The Invasion Episode 6 | 6.5 |
| The Invasion Episode 7 | 7.2 |
| The Invasion Episode 8 | 7 |
| The Krotons Episode 1 | 9 |
| The Krotons Episode 2 | 8.4 |
| The Krotons Episode 3 | 7.5 |
| The Krotons Episode 4 | 7.1 |
| The Seeds of Death Episode 1 | 6.6 |
| The Seeds of Death Episode 2 | 6.8 |
| The Seeds of Death Episode 3 | 7.5 |
| The Seeds of Death Episode 4 | 7.1 |
| The Seeds of Death Episode 5 | 7.6 |
| The Seeds of Death Episode 6 | 7.7 |
| The Space Pirates Episode 1 | 5.8 |
| The Space Pirates Episode 2 | 6.8 |
| The Space Pirates Episode 3 | 6.4 |
| The Space Pirates Episode 4 | 5.8 |
| The Space Pirates Episode 5 | 5.5 |
| The Space Pirates Episode 6 | 5.3 |
| The War Games Episode 1 | 5.5 |
| The War Games Episode 2 | 6.3 |
| The War Games Episode 3 | 5.1 |
| The War Games Episode 4 | 5.7 |
| The War Games Episode 5 | 5.1 |
| The War Games Episode 6 | 4.2 |
| The War Games Episode 7 | 4.9 |
| The War Games Episode 8 | 3.5 |
| The War Games Episode 9 | 4.1 |
| The War Games Episode 10 | 5 |

100%
90%
80%
70%
60%
50%
40%
30%

20
40
60
80
100
120
140
160
180
200
220
240

6
5
4
3
2
1

# Black and White vs Colour

Doctor Who began in 1963, but was only made in colour from Spearhead from Space (1970).

"Kidneys! I've got new kidneys. I don't like the colour."
→ **The Twelfth Doctor, The Time of the Doctor (2013).**

● BLACK AND WHITE EPISODES (30.63%)

● COLOUR EPISODES **(69.37%)**

# The Best of Enemies

The favourite monsters and villains
of the actors who played the Doctor.*

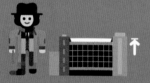

| | | |
|---|---|---|
| **1**<br>DALEKS | **2**<br>ANDROGUMS | **3**<br>OGRONS<br>AND<br>DRACONIANS |
| **4**<br>THE SIXTH FLOOR<br>AT THE BBC | **5**<br>CYBERMEN | **6**<br>THE MASTER |
| **7**<br>DALEKS | **8**<br>YETI | **9**<br>SONTARANS |
| **10**<br>ZYGONS AND<br>SONTARANS | **11**<br>WEEPING<br>ANGELS | **12**<br>MONDASIAN<br>CYBERMEN<br>AND AXONS |

*See page 212 for our references.

35

# One-off Wonders
## Doctors One to Three

Non-human, non-robot creatures featured in one story
of Doctor Who and then never seen – or mentioned – again...*

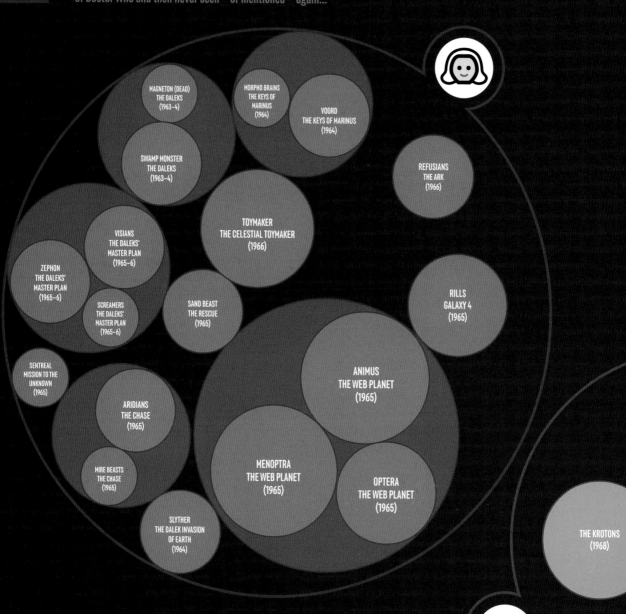

MAGNETON (DEAD)
THE DALEKS
(1963–4)

MORPHO BRAINS
THE KEYS OF
MARINUS
(1964)

VOORD
THE KEYS OF MARINUS
(1964)

SWAMP MONSTER
THE DALEKS
(1963–4)

REFUSIANS
THE ARK
(1966)

VISIANS
THE DALEKS'
MASTER PLAN
(1965–6)

TOYMAKER
THE CELESTIAL TOYMAKER
(1966)

ZEPHON
THE DALEKS'
MASTER PLAN
(1965–6)

SCREAMERS
THE DALEKS'
MASTER PLAN
(1965–6)

SAND BEAST
THE RESCUE
(1965)

RILLS
GALAXY 4
(1965)

SENTREAL
MISSION TO THE
UNKNOWN
(1965)

ANIMUS
THE WEB PLANET
(1965)

ARIDIANS
THE CHASE
(1965)

MIRE BEASTS
THE CHASE
(1965)

MENOPTRA
THE WEB PLANET
(1965)

OPTERA
THE WEB PLANET
(1965)

SLYTHER
THE DALEK INVASION
OF EARTH
(1964)

THE KROTONS
(1968)

SPITTING PLANTS
PLANET OF THE DALEKS
(1973)

SPIRIDONS
PLANET OF THE DALEKS
(1973)

AMBASSADORS
THE AMBASSADORS OF DEATH
(1970)

EXXILONS
DEATH TO THE DALEKS
(1974)

INTER MINORANS
CARNIVAL OF MONSTERS
(1973)

OGRON-EATING
BLOB
FRONTIER IN
SPACE (1973)

UXARIENS
COLONY IN SPACE
(1971)

ARCTURUS
THE CURSE OF PELADON
(1972)

GIANT FLY
THE GREEN DEATH
(1973)

GIANT MAGGOTS
THE GREEN DEATH
(1973)

MIND PARASITE
THE MIND OF EVIL
(1971)

CHRONOVORE
THE TIME MONSTER
(1972)

GEL GUARDS
THE THREE DOCTORS
(1972–3)

GIANT SPIDERS
PLANET OF THE SPIDERS
(1974)

FISH PEOPLE
THE UNDERWATER MENACE
(1967)

THE CHAMELEONS
THE FACELESS ONES
(1967)

WEED CREATURE
FURY FROM THE DEEP
(1968)

*Sized according to the number
of episodes they appear in.

MANDRAGORA HELIX
THE MASQUE OF MANDRAGORA
(1976)

MARSHMEN
FULL CIRCLE
(1980)

MARSHSPIDERS
FULL CIRCLE
(1980)

KASTRIANS
THE HAND OF FEAR
(1976)

THARILS
WARRIORS' GATE
(1981)

GIANT SPIDER
THE TALONS OF
WENG-CHIANG
(1977)

GIANT RAT
THE TALONS OF
WENG-CHIANG
(1977)

SHRIVENZALE
THE RIBOS OPERATION
(1978)

KRAALS
THE ANDROID INVASION
(1975)

OSIRANS
PYRAMIDS OF MARS
(1975)

KROLL
THE POWER OF KROLL
(1978–9)

MEGARA
THE STONES OF BLOOD
(1978)

SWAMPIES
THE POWER OF KROLL
(1978–9)

ANTIMATTER CREATURE
PLANET OF EVIL
(1975)

OGRI
THE STONES OF BLOOD
(1978)

SENTIENT VIRUS
THE INVISIBLE ENEMY
(1977)

JAGAROTH
CITY OF DEATH
(1979)

KRYNOIDS
THE SEEDS OF DOOM
(1976)

FENDAHL
IMAGE OF THE FENDAHL
(1977)

VOGANS
REVENGE OF
THE CYBERMEN
(1975)

USURIANS
THE SUN MAKERS
(1977)

MANDRELS
NIGHTMARE OF EDEN
(1979)

VARDANS
THE INVASION OF TIME
(1978)

BELL PLANTS AND LUSH,
AGGRESSIVE VEGETATION
MEGLOS
(1980)

CARNIVOROUS PLANTS
NIGHTMARE OF EDEN
(1979)

ZOLFA-THURANS
MEGLOS
(1980)

SHADOW
THE ARMAGEDDON FACTOR
(1979)

*Sized according to the number
of episodes they appear in.

# One-off Wonders
## Doctors Four to Five

Non-human, non-robot creatures featured in one story
of Doctor Who and then never seen – or mentioned – again...*

SKARASEN
TERROR OF THE ZYGONS
(1975)

ARGOLIN
THE LEISURE HIVE
(1980)

FOAMASI
THE LEISURE HIVE
(1980)

TARAN WOOD
BEAST
THE ANDROIDS OF
TARA (1978)

TYTHONIANS
THE CREATURE
FROM THE PIT
(1979)

WOLF WEEDS
THE CREATURE
FROM THE PIT
(1979)

HORDA
THE FACE
OF EVIL
(1977)

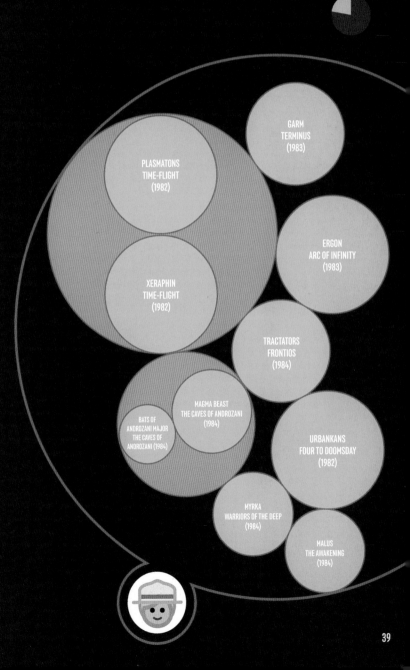

PLASMATONS
TIME-FLIGHT
(1982)

GARM
TERMINUS
(1983)

XERAPHIN
TIME-FLIGHT
(1982)

ERGON
ARC OF INFINITY
(1983)

TRACTATORS
FRONTIOS
(1984)

MAGMA BEAST
THE CAVES OF ANDROZANI
(1984)

BATS OF
ANDROZANI MAJOR
THE CAVES OF
ANDROZANI (1984)

URBANKANS
FOUR TO DOOMSDAY
(1982)

MYRKA
WARRIORS OF THE DEEP
(1984)

MALUS
THE AWAKENING
(1984)

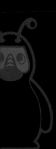

# One-off Wonders
## Doctors Six to Eight

Non-human, non-robot creatures featured in one story
of Doctor Who and then never seen – or mentioned – again...*

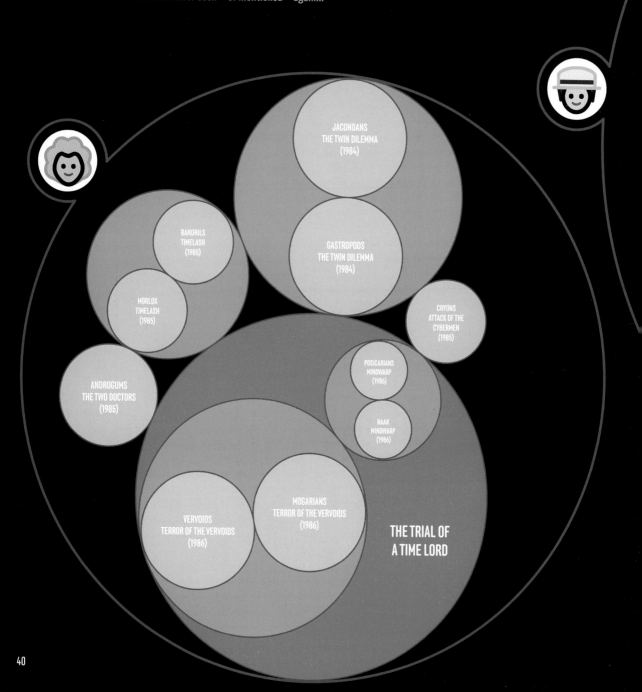

JACONDANS
THE TWIN DILEMMA
(1984)

BANDRILS
TIMELASH
(1985)

GASTROPODS
THE TWIN DILEMMA
(1984)

MORLOX
TIMELASH
(1985)

CRYONS
ATTACK OF THE
CYBERMEN
(1985)

POSICARIANS
MINDWARP
(1986)

ANDROGUMS
THE TWO DOCTORS
(1985)

RAAK
MINDWARP
(1986)

VERVOIDS
TERROR OF THE VERVOIDS
(1986)

MOGARIANS
TERROR OF THE VERVOIDS
(1986)

THE TRIAL OF
A TIME LORD

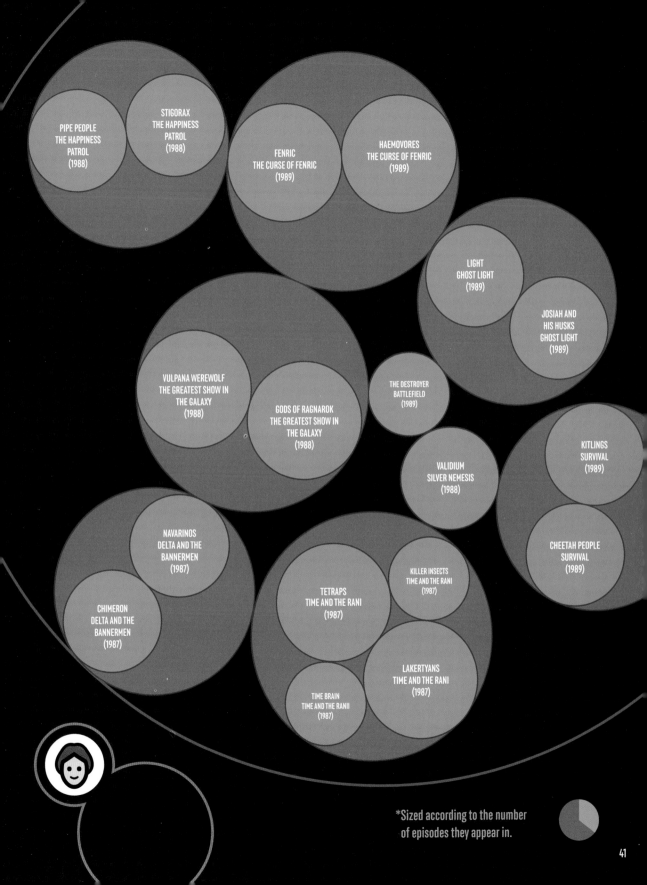

PIPE PEOPLE
THE HAPPINESS
PATROL
(1988)

STIGORAX
THE HAPPINESS
PATROL
(1988)

FENRIC
THE CURSE OF FENRIC
(1989)

HAEMOVORES
THE CURSE OF FENRIC
(1989)

LIGHT
GHOST LIGHT
(1989)

JOSIAH AND
HIS HUSKS
GHOST LIGHT
(1989)

VULPANA WEREWOLF
THE GREATEST SHOW IN
THE GALAXY
(1988)

GODS OF RAGNAROK
THE GREATEST SHOW IN
THE GALAXY
(1988)

THE DESTROYER
BATTLEFIELD
(1989)

KITLINGS
SURVIVAL
(1989)

VALIDIUM
SILVER NEMESIS
(1988)

CHEETAH PEOPLE
SURVIVAL
(1989)

NAVARINOS
DELTA AND THE
BANNERMEN
(1987)

CHIMERON
DELTA AND THE
BANNERMEN
(1987)

TETRAPS
TIME AND THE RANI
(1987)

KILLER INSECTS
TIME AND THE RANI
(1987)

LAKERTYANS
TIME AND THE RANI
(1987)

TIME BRAIN
TIME AND THE RANI
(1987)

*Sized according to the number
of episodes they appear in.

# One-off Wonders
## Doctors Nine to Twelve

Non-human, non-robot creatures featured in one story
of Doctor Who and then never seen – or mentioned – again...*

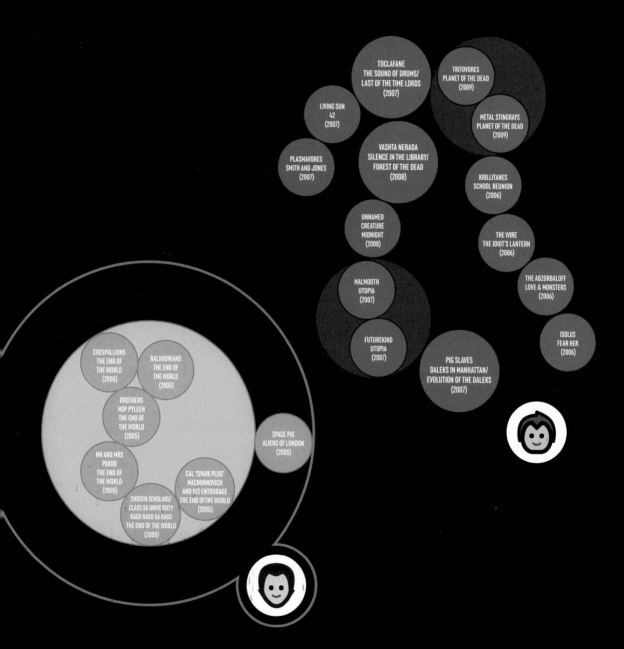

TOCLAFANE
THE SOUND OF DRUMS/
LAST OF THE TIME LORDS
(2007)

TRITOVORES
PLANET OF THE DEAD
(2009)

LIVING SUN
42
(2007)

METAL STINGRAYS
PLANET OF THE DEAD
(2009)

VASHTA NERADA
SILENCE IN THE LIBRARY/
FOREST OF THE DEAD
(2008)

PLASMAVORES
SMITH AND JONES
(2007)

KRILLITANES
SCHOOL REUNION
(2006)

UNNAMED
CREATURE
MIDNIGHT
(2008)

THE WIRE
THE IDIOT'S LANTERN
(2006)

MALMOOTH
UTOPIA
(2007)

THE ABZORBALOFF
LOVE & MONSTERS
(2006)

FUTUREKIND
UTOPIA
(2007)

ISOLUS
FEAR HER
(2006)

PIG SLAVES
DALEKS IN MANHATTAN/
EVOLUTION OF THE DALEKS
(2007)

CRESPALLIONS
THE END OF
THE WORLD
(2005)

BALHOONIANS
THE END OF
THE WORLD
(2005)

BROTHERS
HOP PYLEEN
THE END OF
THE WORLD
(2005)

SPACE PIG
ALIENS OF LONDON
(2005)

MR AND MRS
PAKOO
THE END OF
THE WORLD
(2005)

CAL 'SPARK PLUG'
MACNONNOVICH
AND HIS ENTOURAGE
THE END OF THE WORLD
(2005)

CHOSEN SCHOLARS/
CLASS 55 UNIVERSITY
RAGO RAGO 56 RAGO
THE END OF THE WORLD
(2005)

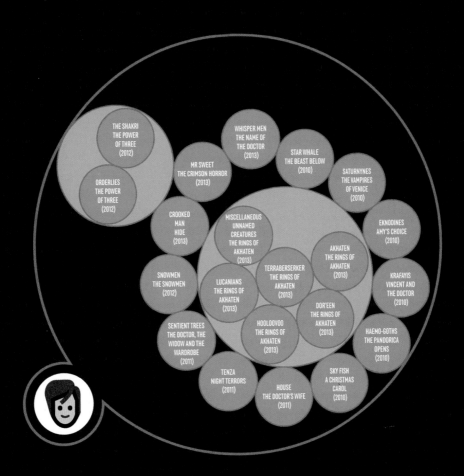

THE SHAKRI
THE POWER
OF THREE
(2012)

ORDERLIES
THE POWER
OF THREE
(2012)

WHISPER MEN
THE NAME OF
THE DOCTOR
(2013)

STAR WHALE
THE BEAST BELOW
(2010)

SATURNYNES
THE VAMPIRES
OF VENICE
(2010)

MR SWEET
THE CRIMSON HORROR
(2013)

CROOKED
MAN
HIDE
(2013)

MISCELLANEOUS
UNNAMED
CREATURES
THE RINGS OF
AKHATEN
(2013)

EKNODINES
AMY'S CHOICE
(2010)

AKHATEN
THE RINGS OF
AKHATEN
(2013)

SNOWMEN
THE SNOWMEN
(2012)

LUCANIANS
THE RINGS OF
AKHATEN
(2013)

TERRABERSERKER
THE RINGS OF
AKHATEN
(2013)

KRAFAYIS
VINCENT AND
THE DOCTOR
(2010)

DOR'EEN
THE RINGS OF
AKHATEN
(2013)

SENTIENT TREES
THE DOCTOR, THE
WIDOW AND THE
WARDROBE
(2011)

HOOLOOVOO
THE RINGS OF
AKHATEN
(2013)

HAEMO-GOTHS
THE PANDORICA
OPENS
(2010)

TENZA
NIGHT TERRORS
(2011)

HOUSE
THE DOCTOR'S WIFE
(2011)

SKY FISH
A CHRISTMAS
CAROL
(2010)

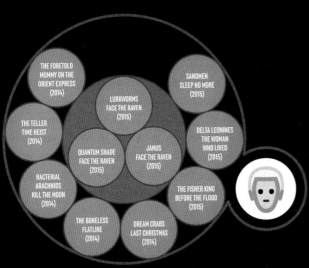

THE FORETOLD
MUMMY ON THE
ORIENT EXPRESS
(2014)

SANDMEN
SLEEP NO MORE
(2015)

LURKWORMS
FACE THE RAVEN
(2015)

THE TELLER
TIME HEIST
(2014)

DELTA LEONINES
THE WOMAN
WHO LIVED
(2015)

QUANTUM SHADE
FACE THE RAVEN
(2015)

JANUS
FACE THE RAVEN
(2015)

BACTERIAL
ARACHNIDS
KILL THE MOON
(2014)

THE FISHER KING
BEFORE THE FLOOD
(2015)

THE BONELESS
FLATLINE
(2014)

DREAM CRABS
LAST CHRISTMAS
(2014)

*Sized according to the number
of episodes they appear in.

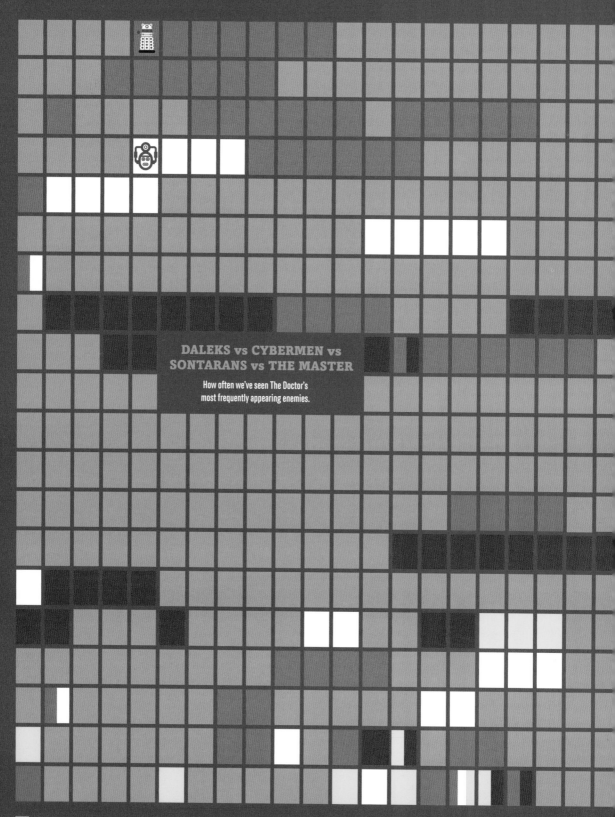

DALEKS vs CYBERMEN vs
SONTARANS vs THE MASTER

How often we've seen The Doctor's
most frequently appearing enemies.

563 episodes without Daleks, Cybermen, Sontarans or The Master

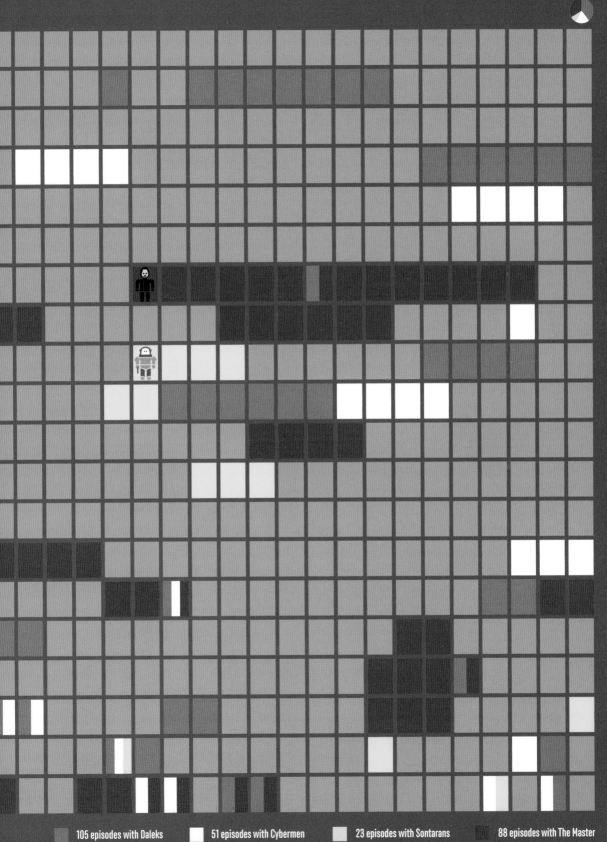

105 episodes with Daleks     51 episodes with Cybermen     23 episodes with Sontarans     88 episodes with The Master

# THE TOMBS OF THE CYBERMEN
### (The Tomb of the Cybermen)
On the planet Telos, archaeologists Klieg and Kaftan awake the Cybermen from their tombs...

WARNING: WILL LIKELY CAUSE NIGHTMARES FOR DECADES TO COME

# THE MASTER ATTACKS BRUCE
### (Doctor Who)
As poor, unsuspecting ambulance-man Bruce sleeps in his bed, the Master, in the guise of a morphant, leaps into his mouth in order to possess him.

WARNING: AVOID IF YOU SUFFER FROM OPHIOPHOBIA

# THE YETI COMES ALIVE
### (The Web of Fear)
After Professor Travers leaves collector Julius Silverstein's house, the Yeti's eyes begin to glow as it comes back to life.

BBFC: CONTAINS SOME MILD PERIL

# SARAH AND THE HAND
### (The Hand of Fear)
Under the control of Eldrad, Sarah watches as a fossilised hand slowly comes back to life.

BBFC: CONTAINS MILD FANTASY HORROR

# MR OAK AND MR QUILL
### (Fury from the Deep)
Creepy duo Mr Oak and Mr Quill enter Maggie Harris' bedroom, opening their mouths as she is overcome by an overwhelming thumping pulse.

WARNING: CONTAINS MIDDLE-AGED MEN IN LIPSTICK

# The Fe
Doctor Who's scariest, eeriest and creepiest moments.

# THE AUTONS BREAK OUT
### (Spearhead from Space)
The Autons awake from their slumber and begin breaking out from shop windows across London, killing all who cross them.

WARNING: CONTAINS EXTREME SHOP VANDALISM

# WHAT'S UNDER THE BED?
### (Listen)
Something is under the young Danny Pink's bed. What is it? Is it a child playing a prank or something more sinister...?

WARNING: CONTAINS MILD AMBIGUITY

# FACELESS POLICEMEN
### (Terror of the Autons)
The Doctor rips away a policeman's plastic face mask, revealing a featureless Auton underneath.

WARNING: CONTAINS FACELESS AUTHORITY FIGURES

# INTO THE MATRIX
### (The Deadly Assassin)
On Gallifrey, the Doctor enters the surreal world of the Matrix, where danger lurks around every corner.

BBFC: CONTAINS MILD VIOLENCE AND HORROR

## TEGAN'S DREAM
(Kinda)

Tegan falls asleep near some windchimes, transporting her into an eerie dream state where she encounters manifestations of her subconscious.

WARNING: SERIOUSLY FREAKY NIGHTMARES

## DOCTOR CONSTANTINE AND THE GAS MASK
(The Empty Child)

As Doctor Constantine explains to the Doctor what is going on in World War II London, he suddenly stiffens up in pain as a gas mask emerges from his face and he begins to ask for his mummy.

BBFC: CONTAINS MILD HORROR

## MR SIN COMES ALIVE
(The Talons of Weng-Chiang)

Opening the dining room door, Leela comes face to face with Mr Sin, a ventriloquist's dummy come to life.

WARNING: CONTAINS MODERATE DUMMY ACTION

## STENGOS BECOMES A DALEK
(Revelation of the Daleks)

Sliding in and out of his Dalek personality, the bodiless head of Natasha's father pleads with her to kill him.

WARNING: CONTAINS SHOTS LIKELY TO MAKE YOU GO 'EUGH!'

"What's wrong with scared? Scared is a superpower."
→ The Twelfth Doctor, Listen

## ar Factor

## THE DEATH OF KANE
(Dragonfire)

Upon realising his planet no longer exists, Kane opens his ship's screen and is burned to death.

WARNING: CONTAINS MILD RAIDERS OF THE LOST ARK REFERENCE

## THE WEEPING ANGELS MOVE IN
(Blink)

Sally Sparrow and Larry Nightingale must get to the TARDIS before the Weeping Angels do, a task that seems easy until the Angels fuse the lights, throwing the room into darkness.

BBFC: CONTAINS MILD FANTASY HORROR

## LYTTON IS TORTURED
(Attack of the Cybermen)

Having captured the mercenary Lytton, the Cybermen torture him in order to discover his plan for blowing up their time vessel, crushing his hands until they're a bloody pulp.

WARNING: BLOOD!

## THE DOCTOR MEETS THE DEVIL
(The Satan Pit)

The Doctor comes face to face with "the Beast", an ancient creature believed to be the basis of the Devil-figure in all the universe's religions.

BBFC: MILD VIOLENCE

## MRS PITT AND THE MUMMY
(Mummy on the Orient Express)

On board the Orient Express, elderly passenger Mrs Pitt is forced to breathe her last by a terrifying creature nobody can see but her.

BBFC: MILD THREAT

# Capture, Escape, Repeat

How often the Second Doctor and his companions are held prisoner in his final story, The War Games.

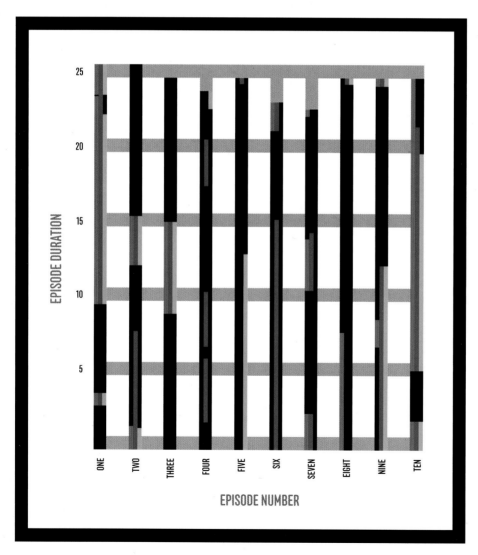

**EPISODE DURATION**

**EPISODE NUMBER**

25

20

15

10

5

ONE TWO THREE FOUR FIVE SIX SEVEN EIGHT NINE TEN

**KEY**

THE DOCTOR

JAMIE

ZOE

"They have escaped before. They could do it again."
→ The Security Chief, The War Games (1969)

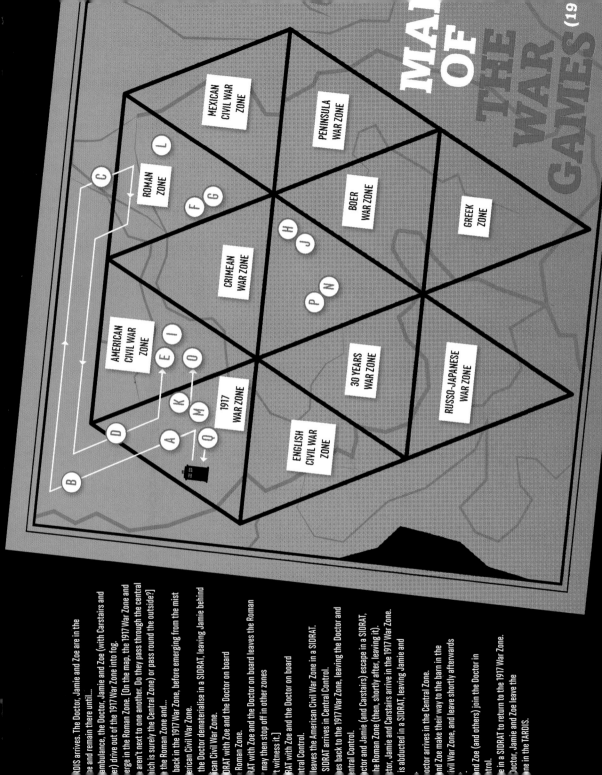

# THE THIRD DOCTOR
## VITAL STATISTICS

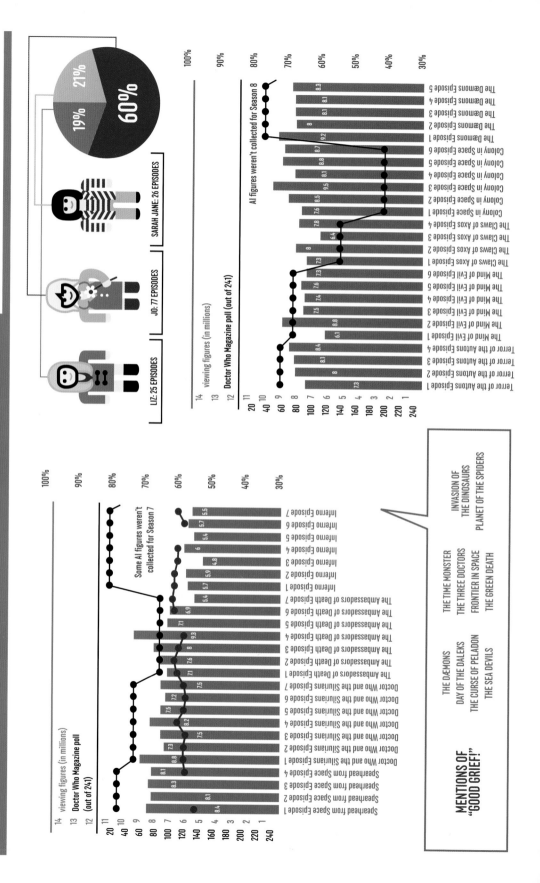

LIZ: 25 EPISODES

JO: 77 EPISODES

SARAH JANE: 26 EPISODES

60%

19%

21%

14 viewing figures (in millions)
13 Doctor Who Magazine poll
12 (out of 241)

**Some AI figures weren't collected for Season 7**

- Spearhead from Space Episode 1
- Spearhead from Space Episode 2 — 8.4, 8.1
- Spearhead from Space Episode 3 — 8.3
- Spearhead from Space Episode 4 — 8.1
- Doctor Who and the Silurians Episode 1 — 8.8
- Doctor Who and the Silurians Episode 2 — 7.3
- Doctor Who and the Silurians Episode 3 — 7.5
- Doctor Who and the Silurians Episode 4 — 8.2
- Doctor Who and the Silurians Episode 5 — 7.2
- Doctor Who and the Silurians Episode 6 — 7.5
- Doctor Who and the Silurians Episode 7 — 7.1
- The Ambassadors of Death Episode 1 — 7.6
- The Ambassadors of Death Episode 2 — 8
- The Ambassadors of Death Episode 3 — 9.3
- The Ambassadors of Death Episode 4
- The Ambassadors of Death Episode 5 — 7.1
- The Ambassadors of Death Episode 6 — 6.9
- The Ambassadors of Death Episode 7 — 5.4
- Inferno Episode 1 — 5.7
- Inferno Episode 2 — 5.9, 4.8
- Inferno Episode 3
- Inferno Episode 4 — 6
- Inferno Episode 5 — 5.4
- Inferno Episode 6 — 5.7
- Inferno Episode 7 — 5.5

**All figures weren't collected for Season 8**

- Terror of the Autons Episode 1 — 7.3
- Terror of the Autons Episode 2 — 8
- Terror of the Autons Episode 3 — 8.1
- Terror of the Autons Episode 4 — 8.4
- The Mind of Evil Episode 1 — 6.1
- The Mind of Evil Episode 2 — 8.8
- The Mind of Evil Episode 3 — 7.5
- The Mind of Evil Episode 4 — 7.4
- The Mind of Evil Episode 5 — 7.6
- The Mind of Evil Episode 6 — 7.3
- The Claws of Axos Episode 1 — 7.3
- The Claws of Axos Episode 2 — 8
- The Claws of Axos Episode 3 — 6.4
- The Claws of Axos Episode 4 — 7.8, 7.6
- Colony in Space Episode 1 — 8.5
- Colony in Space Episode 2 — 9.5
- Colony in Space Episode 3
- Colony in Space Episode 4 — 8.1
- Colony in Space Episode 5 — 8.8
- Colony in Space Episode 6 — 8.7
- The Dæmons Episode 1 — 9.2
- The Dæmons Episode 2 — 8
- The Dæmons Episode 3 — 8.1
- The Dæmons Episode 4 — 8.1
- The Dæmons Episode 5 — 8.3

## MENTIONS OF "GOOD GRIEF!"

THE DÆMONS
DAY OF THE DALEKS
THE CURSE OF PELADON
THE SEA DEVILS

THE TIME MONSTER
THE THREE DOCTORS
FRONTIER IN SPACE
THE GREEN DEATH

INVASION OF THE DINOSAURS
PLANET OF THE SPIDERS

THIRD DOCTOR
::: TARDIS LOG :::

ALTERNATIVE REALITY EARTH
UXARIEUS ▪ PELADON ▪
THROUGH THE BLACK HOLE
TO OMEGA'S DOMAIN ▪ INTER MINOR ▪
EARTH CARGO SHIP C982
UNNAMED OGRON PLANET ▪ SPIRIDON ▪
METEBELIS III ▪ EXXILON

*"Shoes.*
*Must find my shoes."*
FIRST WORDS

MENTIONS OF
"REVERSE THE POLARITY
OF THE NEUTRON FLOW"

THE SEA DEVILS
THE FIVE DOCTORS

VENUSIAN AIKIDO!

THE SEA DEVILS
THE GREEN DEATH
INVASION OF
THE DINOSAURS
INFERNO
THE MIND OF EVIL
THE CLAWS OF AXOS
DAY OF THE DALEKS

SONIC
SCREWDRIVER
MARK II

## Season 10 chart (right)

100%
90%
80%
70%
60%
50%
40%
30%

All figures weren't collected for Season 10

Doctor Who Magazine poll (out of 241)

viewing figures (in millions)

14
13
12

20 11
40 10
60 9
80 8
100 7
120 6
140 5
160 4
180 3
200 2
220 1
240

The Three Doctors Episode 1 — 9.6
The Three Doctors Episode 2 — 10.8
The Three Doctors Episode 3 — 8.8
The Three Doctors Episode 4 — 11.9
Carnival of Monsters Episode 1 — 9.5
Carnival of Monsters Episode 2 — 9
Carnival of Monsters Episode 3 — 9
Carnival of Monsters Episode 4 — 9.2
Frontier in Space Episode 1 — 9.1
Frontier in Space Episode 2 — 7.8
Frontier in Space Episode 3 — 7.5
Frontier in Space Episode 4 — 7.1
Frontier in Space Episode 5 — 7.7
Frontier in Space Episode 6 — 8.9
Planet of the Daleks Episode 1 — 10.7
Planet of the Daleks Episode 2 — 11
Planet of the Daleks Episode 3 — 8.3
Planet of the Daleks Episode 4 — 10.1
Planet of the Daleks Episode 5 — 9.7
Planet of the Daleks Episode 6 — 8.5
The Green Death Episode 1 — 9.2
The Green Death Episode 2 — 7.2
The Green Death Episode 3 — 7.8
The Green Death Episode 4 — 6.8
The Green Death Episode 5 — 8.3
The Green Death Episode 6 — 7

## Season 9 chart (left)

100%
90%
80%
70%
60%
50%
40%
30%

All figures weren't collected for Season 9

Doctor Who Magazine poll (out of 241)

viewing figures (in millions)

14
13
12

20 11
40 10
60 9
80 8
100 7
120 6
140 5
160 4
180 3
200 2
220 1
240

Day of the Daleks Episode 1 — 9.8
Day of the Daleks Episode 2 — 10.4
Day of the Daleks Episode 3 — 9.1
Day of the Daleks Episode 4 — 9.1
The Curse of Peladon Episode 1 — 10.3
The Curse of Peladon Episode 2 — 7.8
The Curse of Peladon Episode 3 — 11
The Curse of Peladon Episode 4 — 8.4
The Sea Devils Episode 1 — 6.4
The Sea Devils Episode 2 — 9.7
The Sea Devils Episode 3 — 8.3
The Sea Devils Episode 4 — 7.8
The Sea Devils Episode 5 — 8.3
The Sea Devils Episode 6 — 8.5
The Mutants Episode 1 — 9.1
The Mutants Episode 2 — 7.9
The Mutants Episode 3 — 7.9
The Mutants Episode 4 — 7.5
The Mutants Episode 5 — 7.9
The Mutants Episode 6 — 6.5
The Time Monster Episode 1 — 7.6
The Time Monster Episode 2 — 7.4
The Time Monster Episode 3 — 8.1
The Time Monster Episode 4 — 7.6
The Time Monster Episode 5 — 6
The Time Monster Episode 6 — 7.6

# THE THIRD DOCTOR
## VITAL STATISTICS

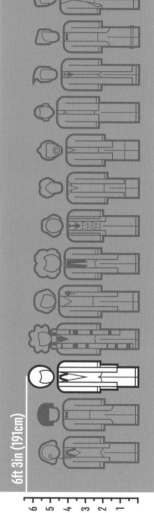

6ft 3in (191cm)

*"A tear,*
*Sarah Jane?*
*No, don't cry.*
*While there's*
*life there's..."*

## LAST WORDS

## FASHION SHOW

**PLANET OF THE DALEKS**

SPEARHEAD FROM SPACE
DOCTOR WHO AND THE SILURIANS
AMBASSADORS OF DEATH
INFERNO
TERROR OF THE AUTONS
THE CLAWS OF AXOS
COLONY IN SPACE
THE SEA DEVILS

TERROR OF THE AUTONS
THE MIND OF EVIL
THE CLAWS OF AXOS
DAY OF THE DAEMONS
THE MUTANTS

THE CURSE OF PELADON
THE TIME MONSTER
THE THREE DOCTORS
PLANET OF THE SPIDERS

CARNIVAL OF MONSTERS
FRONTIER IN SPACE
PLANET OF THE DALEKS
THE GREEN DEATH
THE TIME WARRIOR

**THE GREEN DEATH**
**DEATH TO THE DALEKS**

**INVASION OF THE DINOSAURS**

**THE MONSTER OF PELADON**

**PLANET OF THE SPIDERS**

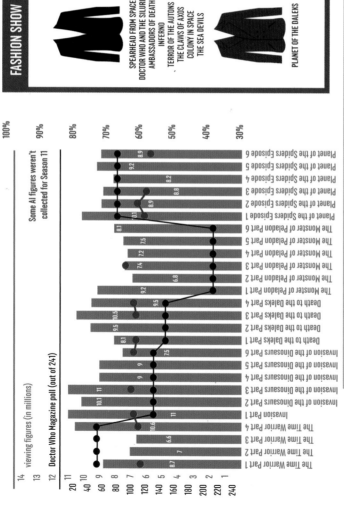

Some AI figures weren't
collected for Season 11

- viewing figures (in millions)
- Doctor Who Magazine poll (out of 241)

100%
90%
80%
70%
60%
50%
40%
30%

20  11
10
40  10
8
60   9
80
100  7
120  6
140  5
160  4
180  3
200  2
220  1
240

The Time Warrior Part 1
The Time Warrior Part 2 — 8.7
The Time Warrior Part 3 — 7
The Time Warrior Part 4 — 6.6
Invasion Part 1 — 11
Invasion of the Dinosaurs Part 2 — 10.4
Invasion of the Dinosaurs Part 3 — 10.1 — 11
Invasion of the Dinosaurs Part 4
Invasion of the Dinosaurs Part 5 — 9 — 9
Invasion of the Dinosaurs Part 6 — 7.5
Death to the Daleks Part 1 — 8.1
Death to the Daleks Part 2 — 9.5
Death to the Daleks Part 3 — 10.5
Death to the Daleks Part 4 — 9.5
The Monster of Peladon Part 1 — 9.2
The Monster of Peladon Part 2 — 6.8
The Monster of Peladon Part 3 — 7.4  7.2
The Monster of Peladon Part 4
The Monster of Peladon Part 5 — 7.5
The Monster of Peladon Part 6 — 8.1
Planet of the Spiders Episode 1 — 10.1
Planet of the Spiders Episode 2 — 8.9
Planet of the Spiders Episode 3 — 8.8
Planet of the Spiders Episode 4 — 8.2
Planet of the Spiders Episode 5 — 9.2
Planet of the Spiders Episode 6 — 8.9

# Structure **of the Earth**

There are beasts below...

**Drowned Racnoss babies**
**6500km miles down in molten core**
*The Runaway Bride*

**Dalek mineshaft from Bedford**
**6.5km from outer core**
*The Dalek Invasion of Earth*

**Primordial slime**
**just below the crust**
*Inferno*

**Silurian civilisation**
**more than 21 km deep**
*Doctor Who and the Silurians*
*Warriors of the Deep*
*The Hungry Earth/ Cold Blood*

**Salamander's atomic deep shelter**
**underneath Australia**
*The Enemy of the World*

"Now, listen to me, all of you!
You are not to attempt to penetrate the Earth's crust!"
→ The Third Doctor, Inferno (1970)

# THE DOCTOR'S EARTHLY TRAVELS

This planet – not just the UK – is protected. Just look at all the places where the TARDIS has been sighted...*

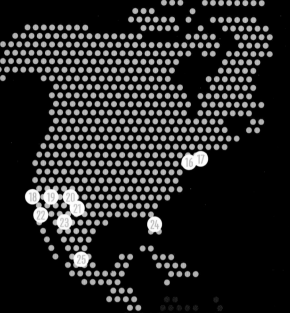

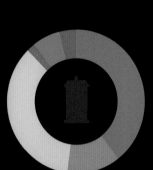

EUROPE **40%**
NORTH AMERICA **35%**
ASIA **13%**
ANTARCTICA **6%**
AFRICA **4%**
AUSTRALASIA **2%**

1  AMSTERDAM, NETHERLANDS – ARC OF INFINITY
2  BERLIN, GERMANY – LET'S KILL HITLER
3  AUVERS-SURS-OISE, FRANCE – VINCENT AND THE DOCTOR
4  POW CAMP, GERMANY – THE IMPOSSIBLE ASTRONAUT
5  PARIS, FRANCE – THE REIGN OF TERROR, THE MASSACRE, CITY OF DEATH,
   THE GIRL IN THE FIREPLACE, VINCENT AND THE DOCTOR
6  VENICE, ITALY – THE VAMPIRES OF VENICE
7  SEA OFF THE AZORES, NORTH ATLANTIC – THE CHASE
8  FLORENCE, ITALY – CITY OF DEATH
9  ROME, ITALY – THE ROMANS
10  POMPEII, ITALY – THE FIRES OF POMPEII
11  SEVILLE, SPAIN – THE TWO DOCTORS
12  SAN MARTINO, ITALY – THE MASQUE OF MANDRAGORA
13  LANZAROTE, CANARY ISLANDS – PLANET OF FIRE
14  NORTH POLAR ICEPACK, ARCTIC OCEAN – COLD WAR
15  MID-ATLANTIC – TIME-FLIGHT
16  WASHINGTON DC, USA – THE IMPOSSIBLE ASTRONAUT/ DAY OF THE MOON
17  NEW YORK CITY, USA – THE CHASE, DALEKS IN MANHATTAN/ EVOLUTION OF THE DALEKS,
   DAY OF THE MOON, THE ANGELS TAKE MANHATTAN
18  SAN FRANCISCO, USA – DOCTOR WHO
19  NEVADA, USA – A TOWN CALLED MERCY, HELL BENT
20  LAKE SILENCIO, UTAH, USA – THE IMPOSSIBLE ASTRONAUT, THE WEDDING OF RIVER SONG
21  UTAH, USA – DALEK
22  LOS ANGELES/ HOLLYWOOD, USA – THE DALEKS' MASTER PLAN
23  TOMBSTONE, ARIZONA, USA – THE GUNFIGHTERS
24  FLORIDA, USA – THE IMPOSSIBLE ASTRONAUT/ DAY OF THE MOON
25  MEXICO – THE AZTECS
26  ATLANTIS, ATLANTIC OCEAN – THE UNDERWATER MENACE, THE TIME MONSTER
27  GOBI DESERT, CHINA/ MONGOLIA – MARCO POLO
28  BEIJING, CHINA – MARCO POLO

*For London sightings, see pages 70–73.

29 ● PALESTINE – THE CRUSADE
30 ● TIBET – THE ABOMINABLE SNOWMEN
31 ● INDIAN SPACE AGENCY, INDIA – DINOSAURS ON A SPACESHIP
32 ● INDIAN OCEAN – THE CURSE OF THE BLACK SPOT
33 ● GIZA, EGYPT – THE DALEKS' MASTER PLAN
34 ● AFRICAN PLAINS, KENYA? – DINOSAURS ON A SPACESHIP
35 ● AUSTRALIA – THE ENEMY OF THE WORLD
36 ● ANTARCTICA – THE SEEDS OF DOOM, THE TENTH PLANET, COLD WAR

UNIT can trace the Doctor's visits to Earth using
an algorithm that generates probabilities,
based on crisis points, anomalies, anachronisms
and keywords such as 'blue box' and 'Doctor'.
→ The Magician's Apprentice (2015)

# A SHORT HISTORY OF THE PLANET EARTH AND HUMANITY...

## PRE-HUMANIAN ERA

**4.6 BILLION YEARS AGO**
Earth begins to form around a Racnoss spaceship, witnessed by the Tenth Doctor and Donna.

**UNKNOWN (BUT AFTER THE PLANET FORMED)**
The Eleventh Doctor and Clara briefly visit the very hot, newly formed Earth.

**C. 400 MILLION YEARS AGO**
A Jagaroth spaceship explodes on Earth's surface, the radiation sparking life in native amniotic fluid, as seen by the Fourth Doctor, Romana and Duggan.

**444–419 MILLION YEARS AGO**
Geological period known as Silurian begins with mass extinction of 60% marine species. Dr Quinn wrongly dates the "Silurian" people to this time.

## HUMANIAN ERA

**12 MILLION YEARS AGO**
Arrival on Earth of the vestiges of the Fendahl, influencing development of Homo sapiens (humans).

**"MILLIONS OF YEARS AGO"**
At an unspecified date, Earth's twin planet Mondas flung out to the edges of the Solar System.

## HISTORICAL PERIOD (AFTER THE INVENTION OF WRITING)

**c. 2,300 BC**
First known visit to Earth by Daleks, in Egypt – as witnessed by the First Doctor.

**c. 1500 BC**
Atlantis destroyed (or submerged) when the Master unleashes the time monster, Kronos.

**1334 BC**
Queen Nefertiti of Egypt becomes the earliest known human to travel in time, in the TARDIS.

**AD 102**
First known visit to Earth by Cybermen, Sontarans and other members of the alliance to trap the Eleventh Doctor in the Pandorica.

## MODERN PERIOD (HUMANS DEFEND EARTH)

**1879**
After an attack by an alien werewolf, Queen Victoria establishes Torchwood to defend Earth from aliens.

**1963**
Two humans, Ian Chesterton and Barbara Wright, kidnapped by the First Doctor – the first (to the Doctor, not chronologically) in a long line of humans to travel in space and time.

**21 JULY 1969**
Neil Armstrong becomes the first human to walk on the Moon; the clip of him doing so is used by the Eleventh Doctor to defeat the Silence.

**c. 1975**
UNIT is established to investigate increasing alien activity on Earth.

## NEAR FUTURE

**2049**
Humans on Earth vote to kill the creature hatching from inside the Moon (its egg) – but Clara overrules them. The Moon hatches, the creature is born and a new egg (or Moon) is left behind.

**NOVEMBER 2059**
The first human colony on Mars, Bowie Base One, is attacked by the Flood. Despite the Tenth Doctor's efforts, all but two of the crew (plus a robot) are killed and Captain Adelaide Brooke kills herself.

**2070**
Humans use a "Gravitron" on the Moon to control Earth's weather.

**c. 2089**
Adelaide Brooke's granddaughter, Susie Fontana Brooke, pilots the first lightspeed ship to Proxima Centauri (the next nearest star after the Sun).

## MIDDLE FUTURE

**LATE 29TH OR EARLY 30TH CENTURY**
Space Station Nerva constructed.

**3000**
Silurians wake from hibernation and peacefully coexist with humans.

## FAR FUTURE

**12,005**
The New Roman Empire, according to the Ninth Doctor.

**UNKNOWN**
At some unknown date, humans settled in megropolises (giant cities) on Pluto have forgotten that their species originated on Earth.

**c. 15,000**
10,000 years after the solar flares, colonists from Earth colonies and humans stored cryogenically on Space Station Nerva return to Earth.

**37,166**
Humanity reaches Zeta Minor, last planet of the known universe.

## POST EARTH

**5,000,000,023**
Humans now living on planet New Earth. Lady Cassandra – the last pure human – is fatally affected (but the Tenth Doctor takes her back in time so she dies in the past).

**100,000,000,000,000**
The Tenth Doctor encounters the last remnants of the human race on Utopia.

## PRE-HUMANIAN ERA

**C. 100 MILLION YEARS AGO**
An unnamed large space creature lays a huge egg in orbit round the Earth.

**>65 MILLION YEARS AGO**
"Silurian" lizard people such as Madame Vastra coexist with dinosaurs. Dinosaurs are first Earth creatures we know of to travel in time: a Tyrannosaurus rex travels to London in the 1890s; various dinosaurs travel to London in the late 1970s.

**65 MILLION YEARS AGO**
The Cybermen crash a spaceship into Earth; dinosaurs become extinct.

**56–33.9 MILLION YEARS AGO**
The Eocene Epoch – during which, the Third Doctor claims, the "Silurian" people flourished.

## HUMANIAN ERA

**500,000 YEARS AGO**
Mankind begins "systematically killing each other", according to the Time Lord known as the War Chief.

**100,000 YEARS AGO**
Dæmons arrive on Earth for the first time and influence human development – as, at some point, do Scaroth of the Jagaroth, the Silence and Osirans. Humans develop fire, the wheel and stone buildings such as the pyramids.

## HISTORICAL PERIOD (AFTER THE INVENTION OF WRITING)

**1066**
Earliest known attempt to alter future Earth history – but the First Doctor stops the Monk from changing the outcome of the Battle of Hastings.

**1562**
First known visit to Earth by Zygons – who are the earliest known refugees of the Time War.

**1866**
First known time travel apparatus built by humans – Professor Theodore Maxtible and Edward Waterfield.

**1869**
In Cardiff, the Ninth Doctor closes a rift running through the middle of the city "like an earthquake fault between different dimensions". A scar remains which creates energy harmless to humans.

## MODERN PERIOD (HUMANS DEFEND EARTH)

**MID-1970S**
Astronauts Charles Carrington and Jim Daniels of Mars Probe 6, encounter aliens on Mars – who accidentally kill Daniels by touching him.

**1978**
Stabiliser failure on his XK-5 spacecraft sends British astronaut Guy Crayford into orbit round Jupiter.

**1986**
Return of Earth's twin planet Mondas – which is then destroyed.

**25 DECEMBER 2006**
First use of a new planetary defence system by Torchwood. It destroys a Sycorax spaceship.

## NEAR FUTURE

**C. END OF 21ST CENTURY**
T-Mat teleportation technology replaces other forms of transport on Earth and to the Moon.

**2157–2167**
Dalek occupation of Earth. The First Doctor foils a Dalek plan to mine out the Earth's core and replace it with a motor.

**2471**
Tens of thousands of people on Earth die every day due to traffic accidents, suicides, pollution and epidemics. Not a blade of grass is left on the planet.

**2526**
Earth hosts conference to unite military forces against Cybermen. Cybermen attempt to disrupt this by crashing a spaceship into Earth – but the Fifth Doctor's companion Adric sends the ship back in time 65 million years.

## MIDDLE FUTURE

**4000**
A human, Mavic Chen, is Guardian of the Galaxy.

**c. 5000**
According to the Fourth Doctor, while Earth is in another Ice Age, the cyborg Peking Homunculus almost causes World War Six. Sometime after this, Earth made uninhabitable by solar flares for c. 10,000 years.

## FAR FUTURE

**c. 200,000**
The Fourth Great and Bountiful Human Empire. Earth is "at its height. Covered with mega-cities, five moons, population 96 billion. The hub of a galactic domain stretching across a million planets, a million species, with mankind right in the middle."

**200,100**
Daleks drop bombs on Europa, Pacifica and the New American Alliance, and destroy Australasia.

**c. 2,000,000,000**
The Time Lords move Earth and its "entire constellation" a couple of light years across space, where – known as Ravalox – it is visited by the Sixth Doctor.

**c. 5,000,000,000**
The Earth is destroyed by the expanding Sun, not quite witnessed by the Ninth Doctor and Rose Tyler, but witnessed on screen by the First Doctor on a spaceship of humans heading for a new home on planet Refusis.

## POST EARTH

**END OF THE UNIVERSE**
Humans Colonel Orson Pink and Clara Oswald visit the last planet left at the end of the universe.

**END OF THE UNIVERSE**
Human hybrid Me (formerly Ashildr) and Clara Oswald witness events on Gallifrey in the last hours of the universe.

# The Political Timeline of **Doctor Who**

British Prime Ministers, according to the series

## WINSTON CHURCHILL
### (1940–1945 / 1951–1955)

Churchill phones the the Eleventh Doctor at the end of The Beast Below and during WWII ordered the creation of "Ironsides" (Daleks) to be used as weapons in the war effort.

## MARGARET THATCHER
### (1979–1990)

The Tenth Doctor confirms that Margaret Thatcher became Prime Minister in 1979. (Tooth and Claw)

## DAVID LLOYD GEORGE
### (1916–1922)

The Ninth Doctor refers to meeting the Liberal Prime Minister in Aliens of London.

## TONY BLAIR
### (1997–200?)

Mickey Smith mentions that Tony Blair has been the Prime Minister. (Rise of the Cybermen)

## "JEREMY"
### (1970s – date unknown)

Thought to be Liberal Leader Jeremy Thorpe, the Prime Minister intervened in the events at Global Chemicals, telling the Brigadier to halt his enquiries. (The Green Death)

**?**

**(2006)**

Prime Minister is killed
by Slitheen and
"Joseph Green" takes
over as Acting PM.
(Aliens of London)

# HAROLD SAXON

## (2008–2009)

...AKA The Master is elected PM.
He is finally shot and
killed by his wife, Lucy.
[History later rewritten.]
(Last of the Time Lords)

# HARRIET JONES

## (2006–2007)

Harriet Jones, MP for Flydale North,
becomes Prime Minister, beginning Britain's
"Golden Age". (The Christmas Invasion)

"Don't you think she looks tired?"
→ The Tenth Doctor,
The Christmas Invasion (2005)

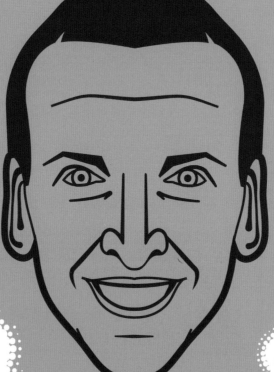

The Empty Child/
The Doctor Dances
(2005)

The Mind
Robber
(1968)

"And everybody lives, Rose!
Everybody lives!
I need more days like this."
→ The Ninth Doctor,
The Doctor Dances (2005)

Fury from
the Deep
(1968)

Hide
(2013)

The Savages
(1966)

Journey to
the Centre of
the TARDIS
(2013)

The Edge of
Destruction
(1964)

Listen
(2014)

# Everybody
# Lives!

**Nobody dies on screen in 8 stories.**

# Decades of The Doctor

When stories have been set – even if for just one scene.

**1900s**
04 — HORROR OF FANG ROCK (1977)

**1910s**
10 — HUMAN NATURE/THE FAMILY OF BLOOD (2007)
04 — PYRAMIDS OF MARS (1975)

**1920s**
10 — THE UNICORN AND THE WASP (2008)
10 — BLINK (2007)
05 — BLACK ORCHID (1982)
03 — CARNIVAL OF MONSTERS (1973)
01 — THE DALEKS' MASTER PLAN (1965–6)

**1930s**
11 — THE ANGELS TAKE MANHATTAN (2012)
11 — LET'S KILL HITLER (2011)
10 — DALEKS IN MANHATTAN/EVOLUTION OF THE DALEKS (2007)
02 — THE ABOMINABLE SNOWMEN (1967)

**1940s**
11 — THE DOCTOR, THE WIDOW AND THE WARDROBE (2011)
11 — THE PANDORICA OPENS (2010)
11 — VICTORY OF THE DALEKS (2010)
11 — THE BEAST BELOW (2010)
09 — THE EMPTY CHILD/THE DOCTOR DANCES (2005)
07 — THE CURSE OF FENRIC (1989)

**1950s**
10 — THE IDIOT'S LANTERN (2006)
07 — DELTA AND THE BANNERMEN (1987)

"I've had decades to think nice thoughts about [the Doctor]. Got a bit harder to stay charitable once I entered decade four."
→ Amy Pond, The Girl Who Waited (2011)

1960s – 2010s →

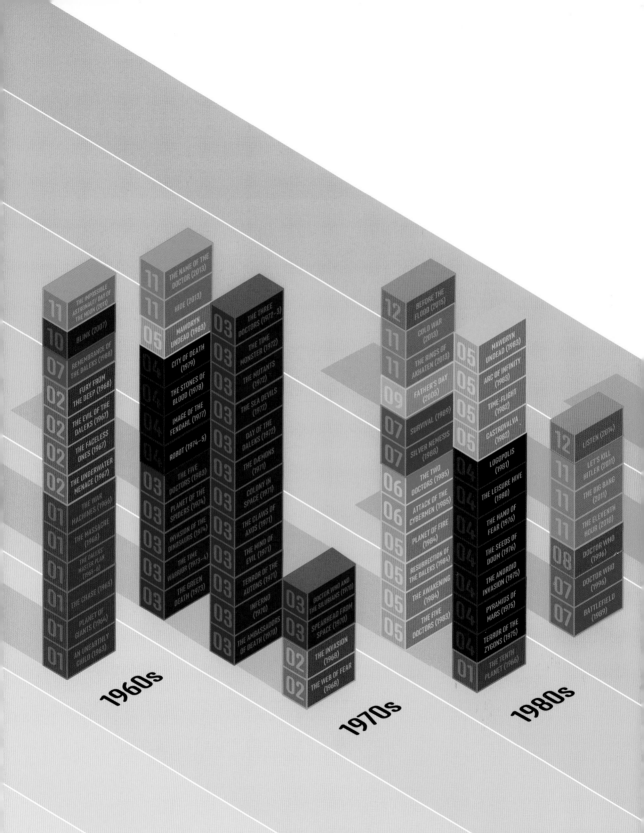

**1960s**

11 — THE IMPOSSIBLE ASTRONAUT/ DAY OF THE MOON (2011)
10 — BLINK (2007)
07 — REMEMBRANCE OF THE DALEKS (1988)
02 — FURY FROM THE DEEP (1968)
02 — THE EVIL OF THE DALEKS (1967)
02 — THE FACELESS ONES (1967)
02 — THE UNDERWATER MENACE (1967)
01 — THE WAR MACHINES (1966)
01 — THE MASSACRE (1966)
01 — THE DALEKS' MASTER PLAN (1965-6)
01 — THE CHASE (1965)
01 — PLANET OF GIANTS (1964)
01 — AN UNEARTHLY CHILD (1963)

11 — THE NAME OF THE DOCTOR (2013)
11 — HIDE (2013)
05 — MAWDRYN UNDEAD (1983)
04 — CITY OF DEATH (1979)
04 — THE STONES OF BLOOD (1978)
04 — IMAGE OF THE FENDAHL (1977)
04 — ROBOT (1974-5)
03 — THE FIVE DOCTORS (1983)
03 — PLANET OF THE SPIDERS (1974)
03 — INVASION OF THE DINOSAURS (1974)
03 — THE TIME WARRIOR (1973-4)
03 — THE GREEN DEATH (1973)

03 — THE THREE DOCTORS (1972-3)
03 — THE TIME MONSTER (1972)
03 — THE MUTANTS (1972)
03 — THE SEA DEVILS (1972)
03 — DAY OF THE DALEKS (1972)
03 — THE DÆMONS (1971)
03 — COLONY IN SPACE (1971)
03 — THE CLAWS OF AXOS (1971)
03 — THE MIND OF EVIL (1971)
03 — TERROR OF THE AUTONS (1971)
03 — INFERNO (1970)
03 — THE AMBASSADORS OF DEATH (1970)

03 — DOCTOR WHO AND THE SILURIANS (1970)
03 — SPEARHEAD FROM SPACE (1970)
02 — THE INVASION (1968)
02 — THE WEB OF FEAR (1968)

**1970s**

12 — BEFORE THE FLOOD (2015)
11 — COLD WAR (2013)
11 — THE RINGS OF AKHATEN (2013)
09 — FATHER'S DAY (2005)
07 — SURVIVAL (1989)
07 — SILVER NEMESIS (1988)
06 — THE TWO DOCTORS (1985)
06 — ATTACK OF THE CYBERMEN (1985)
05 — PLANET OF FIRE (1984)
05 — RESURRECTION OF THE DALEKS (1984)
05 — THE AWAKENING (1984)
05 — THE FIVE DOCTORS (1983)

05 — MAWDRYN UNDEAD (1983)
05 — ARC OF INFINITY (1983)
05 — TIME-FLIGHT (1982)
05 — CASTROVALVA (1982)
04 — LOGOPOLIS (1981)
04 — THE LEISURE HIVE (1980)
04 — THE HAND OF FEAR (1976)
04 — THE SEEDS OF DOOM (1976)
04 — THE ANDROID INVASION (1975)
04 — PYRAMIDS OF MARS (1975)
04 — TERROR OF THE ZYGONS (1975)
01 — THE TENTH PLANET (1966)

**1980s**

12 — LISTEN (2014)
11 — LET'S KILL HITLER (2011)
11 — THE BIG BANG (2011)
11 — THE ELEVENTH HOUR (2010)
08 — DOCTOR WHO (1996)
07 — DOCTOR WHO (1996)
07 — BATTLEFIELD (1989)

64

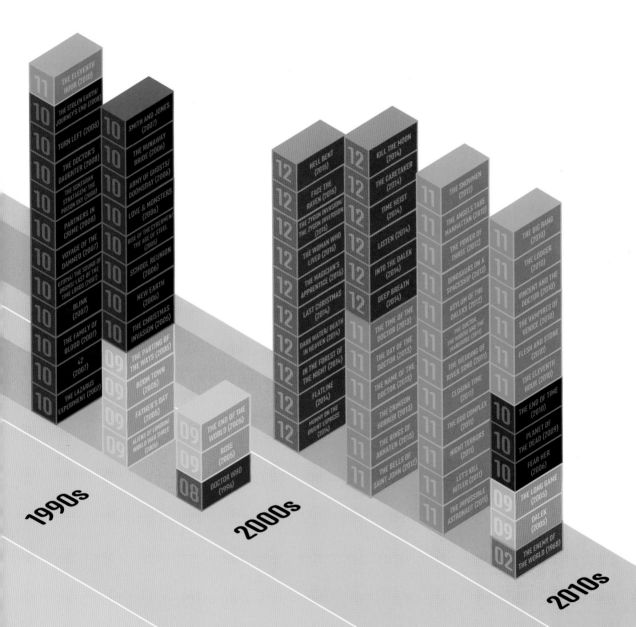

1990s

2000s

2010s

**THE ELEVENTH HOUR (2010)** 11

**THE STOLEN EARTH/ JOURNEY'S END (2008)** 10

**TURN LEFT (2008)** 10

**THE DOCTOR'S DAUGHTER (2008)** 10

**THE SONTARAN STRATAGEM/ THE POISON SKY (2008)** 10

**PARTNERS IN CRIME (2008)** 10

**VOYAGE OF THE DAMNED (2007)** 10

**UTOPIA/ THE SOUND OF DRUMS / LAST OF THE TIME LORDS (2007)** 10

**BLINK (2007)** 10

**THE FAMILY OF BLOOD (2007)** 10

**42 (2007)** 10

**THE LAZARUS EXPERIMENT (2007)** 10

**SMITH AND JONES (2007)** 10

**THE RUNAWAY BRIDE (2006)** 10

**ARMY OF GHOSTS/ DOOMSDAY (2006)** 10

**LOVE & MONSTERS (2006)** 10

**RISE OF THE CYBERMEN/ THE AGE OF STEEL (2005)** 10

**SCHOOL REUNION (2006)** 10

**NEW EARTH (2006)** 10

**THE CHRISTMAS INVASION (2005)** 10

**THE PARTING OF THE WAYS (2005)** 09

**BOOM TOWN (2005)** 09

**FATHER'S DAY (2005)** 09

**ALIENS OF LONDON/ WORLD WAR THREE (2005)** 09

**THE END OF THE WORLD (2005)** 09

**ROSE (2005)** 09

**DOCTOR WHO (1996)** 08

**HELL BENT (2015)** 12

**FACE THE RAVEN (2015)** 12

**THE ZYGON INVASION/ THE ZYGON INVERSION (2015)** 12

**THE WOMAN WHO LIVED (2015)** 12

**THE MAGICIAN'S APPRENTICE (2015)** 12

**LAST CHRISTMAS (2014)** 12

**DARK WATER/ DEATH IN HEAVEN (2014)** 12

**IN THE FOREST OF THE NIGHT (2014)** 12

**FLATLINE (2014)** 12

**MUMMY ON THE ORIENT EXPRESS (2014)** 12

**KILL THE MOON (2014)** 12

**THE CARETAKER (2014)** 12

**TIME HEIST (2014)** 12

**LISTEN (2014)** 12

**INTO THE DALEK (2014)** 12

**DEEP BREATH (2014)** 12

**THE TIME OF THE DOCTOR (2013)** 11

**THE DAY OF THE DOCTOR (2013)** 11

**THE NAME OF THE DOCTOR (2013)** 11

**THE CRIMSON HORROR (2013)** 11

**THE RINGS OF AKHATEN (2013)** 11

**THE BELLS OF SAINT JOHN (2012)** 11

**THE SNOWMEN (2012)** 11

**THE ANGELS TAKE MANHATTAN (2012)** 11

**THE POWER OF THREE (2012)** 11

**DINOSAURS ON A SPACESHIP (2012)** 11

**ASYLUM OF THE DALEKS (2012)** 11

**THE DOCTOR, THE WIDOW AND THE WARDROBE (2011)** 11

**THE WEDDING OF RIVER SONG (2011)** 11

**CLOSING TIME (2011)** 11

**THE GOD COMPLEX (2011)** 11

**NIGHT TERRORS (2011)** 11

**LET'S KILL HITLER (2011)** 11

**THE IMPOSSIBLE ASTRONAUT (2011)** 11

**THE BIG BANG (2010)** 11

**THE LODGER (2010)** 11

**VINCENT AND THE DOCTOR (2010)** 11

**THE VAMPIRES OF VENICE (2010)** 11

**FLESH AND STONE (2010)** 11

**THE ELEVENTH HOUR (2010)** 11

**THE END OF TIME (2010)** 10

**PLANET OF THE DEAD (2009)** 10

**FEAR HER (2006)** 10

**THE LONG GAME (2005)** 09

**DALEK (2005)** 09

**THE ENEMY OF THE WORLD (1968)** 02

65

"I'm the Doctor, and if there's one thing I can do, it's talk. I've got five billion languages, and you haven't got one way of stopping me."
→ The Ninth Doctor, The Parting of the Ways (2005)

# THE LANGUAGES OF THE DOCTOR

BABY • HORSE • CAT • DINOSAUR

The five episodes of Doctor Who
at least partly set on the day they were broadcast.

# The Master
## Through Time

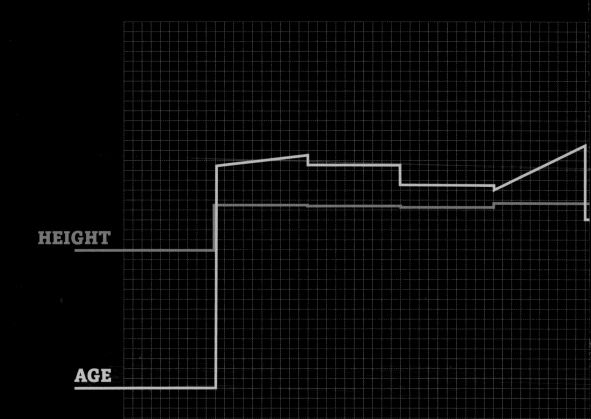

HEIGHT

AGE

Outward appearance
of age, taking the
actors' ages as guidance

| PLAYED BY | ROGER DELGADO | PETER PRATT | GEOFFREY BEEVERS | ANTHONY AINLEY |

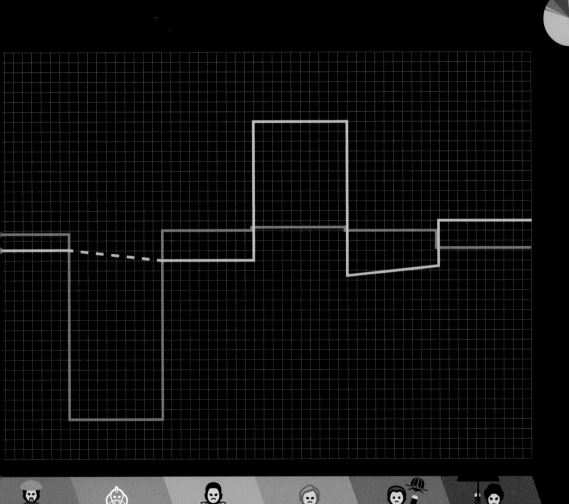

GORDON
TIPPLE
(1996)

MORPHANT
SNAKE
(1996)

ERIC ROBERTS
(1996)

DEREK JACOBI
(2007)

JOHN SIMM
(2007–2010)

MICHELLE GOMEZ
(2014– )

# CALLING THE CAPITAL

Doctor Who stories with at least one scene set in London.

## PART ONE: 1963–1989

(WHEN THE SERIES WAS BEING MADE IN LONDON)

**OVERVIEW**

1 ● SHOREDITCH – AN UNEARTHLY CHILD (1963), ATTACK OF THE CYBERMEN (1985), REMEMBRANCE OF THE DALEKS (1988)

2 ● PIMLICO (GROSVENOR ROAD) – THE DALEK INVASION OF EARTH (1964)

3 ● SLOANE SQUARE – THE DALEK INVASION OF EARTH (1964)

4 ● WHITEHALL – THE DALEK INVASION OF EARTH (1964), INVASION OF THE DINOSAURS (1974)

5 ● TRAFALGAR SQUARE – THE DALEK INVASION OF EARTH (1964), THE CHASE (1965), THE DALEKS' MASTER PLAN (1965–1966), INVASION OF THE DINOSAURS (1974), **THE SONTARAN EXPERIMENT (1975)**

6 ● ALBERT MEMORIAL & ALBERT HALL – THE DALEK INVASION OF EARTH (1964)

7 ● WESTMINSTER BRIDGE – THE DALEK INVASION OF EARTH (1964), INVASION OF THE DINOSAURS (1974)

8 ● KENSINGTON GARDENS – THE CHASE (1965)

9 ● BAYSWATER ROAD – THE CHASE (1965)

10 ● REGENT STREET – THE CHASE (1965)

11 ● OVAL CRICKET GROUND – THE DALEKS' MASTER PLAN (1966)

12 ● WINDMILL ROAD, WIMBLEDON – THE MASSACRE (1966)

13 ● POST OFFICE TOWER – THE WAR MACHINES (1966)

14 ● FITZROY SQUARE GARDENS – THE WAR MACHINES (1966)

15 ● ROYAL OPERA HOUSE, COVENT GARDEN – THE WAR MACHINES (1966)

16 ● COVENT GARDEN MARKET – THE WAR MACHINES (1966), THE WEB OF FEAR (1968), INVASION OF THE DINOSAURS (1974)

17 ● CORNWALL GARDENS WALK – THE WAR MACHINES (1966)

18 ● GATWICK AIRPORT – THE FACELESS ONES (1967), THE EVIL OF THE DALEKS (1967)

19 ● KING'S CROSS UNDERGROUND STATION – THE WEB OF FEAR (1968)

20 ● GOODGE STREET UNDERGROUND STATION – THE WEB OF FEAR (1968)

21 ● HOLBORN UNDERGROUND STATION – THE WEB OF FEAR (1968)

22 ● ST PAUL'S UNDERGROUND STATION – THE WEB OF FEAR (1968)

23 ● COVENT GARDEN UNDERGROUND STATION – THE WEB OF FEAR (1968)

24 ● PICCADILLY CIRCUS UNDERGROUND STATION – THE WEB OF FEAR (1968)

25 ● CHARING CROSS UNDERGROUND STATION – THE WEB OF FEAR (1968)

26 ● MONUMENT UNDERGROUND STATION – THE WEB OF FEAR (1968)

27 ● NEAL STREET AND SHELTON STREET, SOHO – THE WEB OF FEAR (1968)

28 ● 1 MILLBANK – THE INVASION (1968), **TERROR OF THE ZYGONS (1975)**

29 ● REGENT'S CANAL, MAIDA VALE – THE INVASION (1968)

30 ● ST PETER'S STEPS – THE INVASION (1968)

31 ● EALING BROADWAY – SPEARHEAD FROM SPACE (1970)

32 ● MARYLEBONE STATION – DOCTOR WHO AND THE SILURIANS (1970)

33 ● CORNWALL GARDENS – THE MIND OF EVIL (1971)

34 ● SMITHFIELD MARKET – INVASION OF THE DINOSAURS (1974)

35 ● MOORGATE STATION – INVASION OF THE DINOSAURS (1974)

36 ● ALBERT EMBANKMENT – INVASION OF THE DINOSAURS (1974)

37 ● HAYMARKET – INVASION OF THE DINOSAURS (1974)

38 ● OLD BILLINGSGATE MARKET – INVASION OF THE DINOSAURS (1974)

39 ● PICCADILLY – THE SONTARAN EXPERIMENT (1975)

40 ● NORTH SIDE OF BLACKFRIARS BRIDGE – THE TALONS OF WENG-CHIANG (1977)

41 ● ST KATHERINE'S DOCK – THE TALONS OF WENG-CHIANG (1977)

42 ● BETWEEN WHITECHAPEL AND ST GEORGE'S – THE TALONS OF WENG-CHIANG (1977)

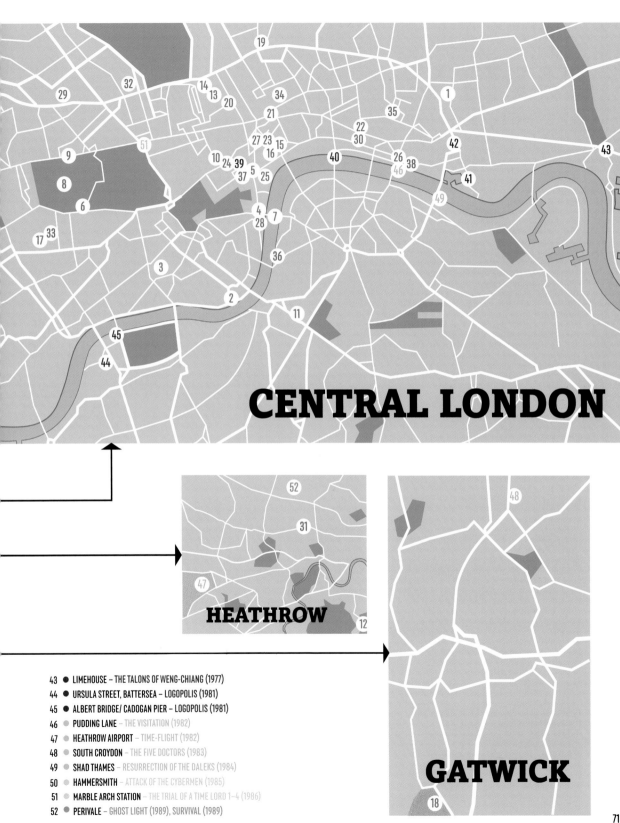

# CENTRAL LONDON

## HEATHROW

## GATWICK

# CALLING THE CAPITAL

## PART TWO: 2005–2015
(WHEN THE SERIES WAS BEING MADE IN CARDIFF)

Doctor Who stories with at least one scene set in an identifiable* part of London.

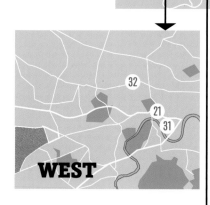

**OVERVIEW**

**WEST**

**EAST**

1 ● **PICCADILLY CIRCUS** – ROSE (2005)

2 ● **TRAFALGAR SQUARE** – ROSE (2005), THE DAY OF THE DOCTOR (2013), IN THE FOREST OF THE NIGHT (2014)

3 ● **EMBANKMENT → ROYAL AIR FORCE MEMORIAL → OVER WESTMINSTER BRIDGE → TO THE LONDON EYE** – ROSE (2005)

4 ● **KENNINGTON** – ROSE (2005), ALIENS OF LONDON/ WORLD WAR THREE (2005), FATHER'S DAY (2005), THE EMPTY CHILD/ THE DOCTOR DANCES (2005), THE PARTING OF THE WAYS (2005), **THE CHRISTMAS INVASION (2005), NEW EARTH (2006), RISE OF THE CYBERMEN/ THE AGE OF STEEL (2006), LOVE & MONSTERS (2006), ARMY OF GHOSTS/ DOOMSDAY (2006), THE END OF TIME (2010)**

5 ● **10 DOWNING STREET** – ALIENS OF LONDON/ WORLD WAR THREE (2005), THE CHRISTMAS INVASION (2005), THE SOUND OF DRUMS/ LAST OF THE TIME LORDS (2007)

6 ● **THE END OF WHITEHALL** – ALIENS OF LONDON/ WORLD WAR THREE (2005)

7 ● **FLIGHT OF THE SPACE PIG** – ALIENS OF LONDON/ WORLD WAR THREE (2005)

8 ● **ST PAUL'S CATHEDRAL** – THE EMPTY CHILD/ THE DOCTOR DANCES (2005), DARK WATER/ DEATH IN HEAVEN (2014)

9 ● **THE PALACE OF WESTMINSTER** – ALIENS OF LONDON (2005), THE EMPTY CHILD/ THE DOCTOR DANCES (2005), **THE ELEVENTH HOUR (2010), DEEP BREATH (2014)**

10 ● **LIMEHOUSE GREEN** – THE EMPTY CHILD/ THE DOCTOR DANCES (2005)

11 ● **THE TOWER OF LONDON** – THE CHRISTMAS INVASION (2005), THE POWER OF THREE (2012), THE DAY OF THE DOCTOR (2013), THE MAGICIAN'S APPRENTICE (2015), THE ZYGON INVASION / THE ZYGON INVERSION (2015)

12 ● **30 ST MARY AXE** – THE CHRISTMAS INVASION (2005)

13 ● **LAMBETH PIER** – RISE OF THE CYBERMEN/ THE AGE OF STEEL (2006)

14 ● **BATTERSEA POWER STATION** – RISE OF THE CYBERMEN/ THE AGE OF STEEL (2006)

15 ● **CHURCHILL GARDENS** – RISE OF THE CYBERMEN/ THE AGE OF STEEL (2006)

16 ● **ALEXANDRA PALACE AND MUSWELL HILL** – THE IDIOT'S LANTERN (2006)

17 ● **WOOLWICH** – LOVE & MONSTERS (2006)

18 ● **STRATFORD** – FEAR HER (2006)

19 ● **ONE CANADA SQUARE, CANARY WHARF** – ARMY OF GHOSTS/ DOOMSDAY (2006)

20 ● **WESTMINSTER BRIDGE** – ARMY OF GHOSTS/ DOOMSDAY (2006), THE BELLS OF SAINT JOHN (2013), DEEP BREATH (2014)

21 ● **CHISWICK FLYOVER/ M4 AND GREAT WEST ROAD** – THE RUNAWAY BRIDE (2006)

22 ● **INTERNATIONAL PRESS CENTRE** – THE RUNAWAY BRIDE (2006)

23 ● **THAMES BARRIER** – THE RUNAWAY BRIDE (2006)

24 ● **ROYAL HOPE HOSPITAL (IN REALITY ST THOMAS'S HOSPITAL)** – SMITH AND JONES (2007)

25 ● **SOUTHWARK** – THE SHAKESPEARE CODE (2007)

26 ● **THE GLOBE THEATRE** – THE SHAKESPEARE CODE (2007)

27 ● **BEDLAM (NOW LIVERPOOL STREET STATION)** – THE SHAKESPEARE CODE (2007)

28 ● **SOUTHWARK CATHEDRAL** – THE LAZARUS EXPERIMENT (2007)

29 ● **COLLEGE GREEN** – THE SOUND OF DRUMS/ LAST OF THE TIME LORDS (2007)

30 ● **BUCKINGHAM PALACE** – VOYAGE OF THE DAMNED (2007), THE WEDDING OF RIVER SONG (2011)

31 ● **CHISWICK** – PARTNERS IN CRIME (2008), THE SONTARAN STRATAGEM/ THE POISON SKY (2008), TURN LEFT (2008), THE STOLEN EARTH/ JOURNEY'S END (2008), THE END OF TIME (2009–10)

32 ● **EALING** – THE STOLEN EARTH/ JOURNEY'S END (2008), THE END OF TIME (2009–10)

33 ● **THE CYBERKING'S ROUTE** – THE NEXT DOCTOR (2008)

34 ● **BETWEEN VICTORIA AND BRIXTON** – PLANET OF THE DEAD (2009)

35 ● **CHURCHILL WAR ROOMS** – THE BEAST BELOW (2010), VICTORY OF THE DALEKS (2010)

36 ● **PATERNOSTER ROW** – A GOOD MAN GOES TO WAR (2011), THE SNOWMEN (2012), THE NAME OF THE DOCTOR (2013), DEEP BREATH (2014)

# CENTRAL LONDON

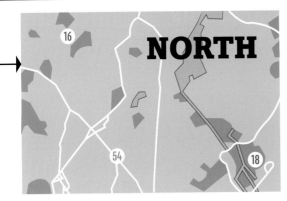

# NORTH

# SOUTH

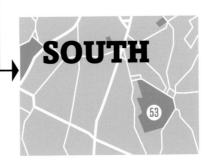

46 ● SHOREDITCH – THE DAY OF THE DOCTOR (2013), INTO THE DALEK (2014), LISTEN (2014), THE CARETAKER (2014), KILL THE MOON (2014), IN THE FOREST OF THE NIGHT (2014), THE MAGICIAN'S APPRENTICE (2015), THE WOMAN WHO LIVED (2015)

47 ● RIVERBANK OPPOSITE THE PALACE OF WESTMINSTER – DEEP BREATH (2014)

48 ● EXHIBITION ROAD – IN THE FOREST OF THE NIGHT (2014)

49 ● KNIGHTSBRIDGE UNDERGROUND STATION – IN THE FOREST OF THE NIGHT (2014)

50 ● NORTHUMBERLAND AVENUE – IN THE FOREST OF THE NIGHT (2014)

51 ● BROMPTON ROAD – IN THE FOREST OF THE NIGHT (2014)

52 ● ST PETER'S STEPS – IN THE FOREST OF THE NIGHT (2014)

53 ● BROCKWELL PARK – THE ZYGON INVASION / THE ZYGON INVERSION (2015)

54 ● HIGHBURY CORNER – FACE THE RAVEN (2015)

37 ● SAVOY HOTEL, THE STRAND – THE POWER OF THREE (2012)

38 ● SKY BAR, ST PAUL'S HOTEL – THE BELLS OF SAINT JOHN (2013)

39 ● WATERLOO BRIDGE – THE BELLS OF SAINT JOHN (2013)

40 ● HORSE GUARDS PARADE – THE BELLS OF SAINT JOHN (2013)

41 ● ADMIRALTY ARCH – THE BELLS OF SAINT JOHN (2013)

42 ● THE QUEEN'S WALK – THE BELLS OF SAINT JOHN (2013)

43 ● ST THOMAS STREET – THE BELLS OF SAINT JOHN (2013)

44 ● THE SHARD – THE BELLS OF SAINT JOHN (2013)

45 ● FULHAM – THE BELLS OF SAINT JOHN (2013), THE RINGS OF AKHATEN (2013) THE CRIMSON HORROR (2013), THE NAME OF THE DOCTOR (2013)

* Many stories feature scenes set in London without stating exactly where they take place.

# Poop Poop!

All the vehicles driven or piloted by the Third Doctor
— and the stories they first appeared in.

RECOVERY 7 CAPSULE

"BESSIE"
– SIVA EDWARDIAN TOURER KIT
FITTED ON E93A FORD POPULAR
CHASSIS, REGISTRATION WHO 1

LOW LOADER

1927 VAUXHALL 14/40
PRINCETON TOURER,
REGISTRATION NF 3226

WHEELCHAIR

SPEARHEAD FROM SPACE

DOCTOR WHO AND
THE SILURIANS

THE AMBASSADORS OF DEATH

THE TARDIS

POLICE VAN

THE TARDIS CONSOLE

"CHARLIE"
– IMC BUGGY

INFERNO

THE MIND
OF EVIL

THE CLAWS
OF AXOS

COLONY IN
SPACE

UNIT LAND ROVER,
REGISTRATION 02BH41

MOTORBIKE

TRIKE

THE DÆMONS

DAY OF THE DALEKS

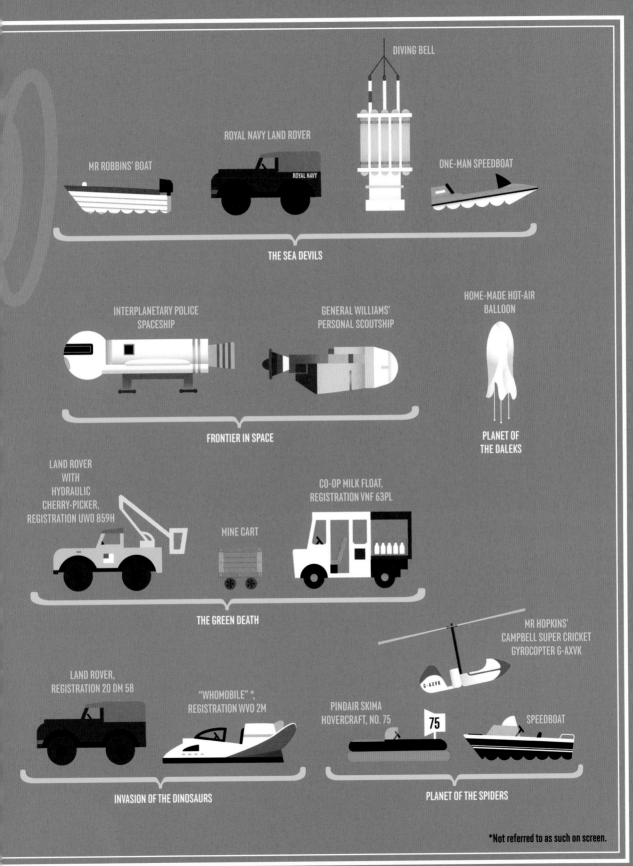

DIVING BELL

ROYAL NAVY LAND ROVER

MR ROBBINS' BOAT

ONE-MAN SPEEDBOAT

ROYAL NAVY

THE SEA DEVILS

INTERPLANETARY POLICE
SPACESHIP

GENERAL WILLIAMS'
PERSONAL SCOUTSHIP

HOME-MADE HOT-AIR
BALLOON

FRONTIER IN SPACE

PLANET OF
THE DALEKS

LAND ROVER
WITH
HYDRAULIC
CHERRY-PICKER,
REGISTRATION UWO 859H

MINE CART

CO-OP MILK FLOAT,
REGISTRATION VNF 63PL

THE GREEN DEATH

MR HOPKINS'
CAMPBELL SUPER CRICKET
GYROCOPTER G-AXVK

G-AXVK

LAND ROVER,
REGISTRATION 20 DM 58

"WHOMOBILE" *,
REGISTRATION WVO 2M

PINDAIR SKIMA
HOVERCRAFT, NO. 75

75

SPEEDBOAT

INVASION OF THE DINOSAURS

PLANET OF THE SPIDERS

*Not referred to as such on screen.

# CHAPTER FOUR "THAT'S SILLY"

## CHAPTER NINE
### DALEKS

## CHAPTER SEVEN
### TECHNOLOGY

## CHAPTER ELEVEN
### ALIEN WORLDS

APPENDIX
THE MAKING OF WHOGRAPHICA

CHAPTER EIGHT
MIND THE GAP

## CHAPTER TWELVE
### 16 DAYS, 8 HOURS, 13 MINUTES, 51 SECONDS... AND COUNTING

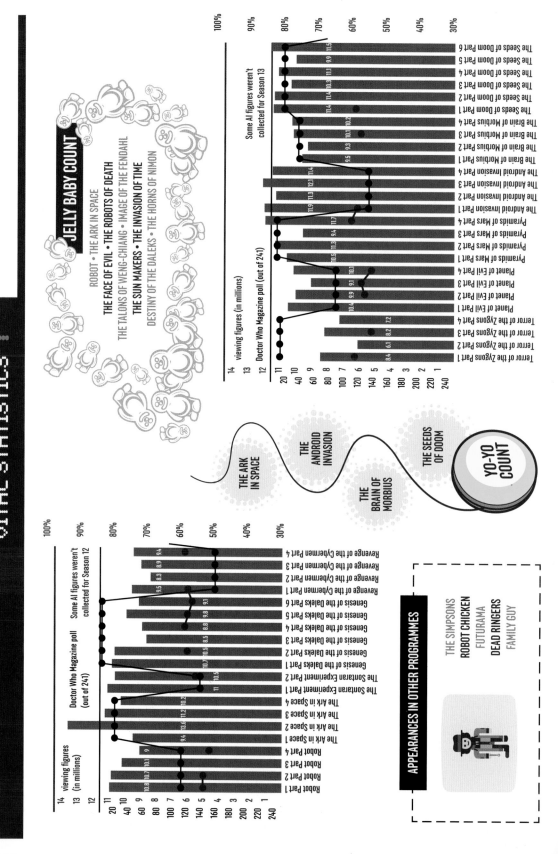

# THE FOURTH DOCTOR
## VITAL STATISTICS

**JELLY BABY COUNT**

ROBOT • THE ARK IN SPACE
**THE FACE OF EVIL • THE ROBOTS OF DEATH**
THE TALONS OF WENG-CHIANG • IMAGE OF THE FENDAHL
**THE SUN MAKERS • THE INVASION OF TIME**
DESTINY OF THE DALEKS • THE HORNS OF NIMON

**YO-YO COUNT**

THE ARK IN SPACE
THE ANDROID INVASION
THE BRAIN OF MORBIUS
THE SEEDS OF DOOM

**APPEARANCES IN OTHER PROGRAMMES**

THE SIMPSONS
ROBOT CHICKEN
FUTURAMA
DEAD RINGERS
FAMILY GUY

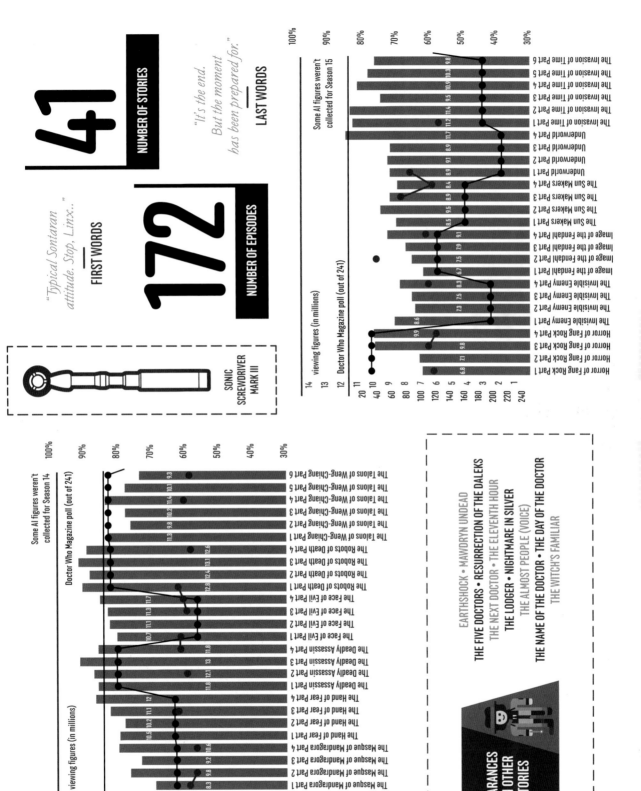

# 41
**NUMBER OF STORIES**

# 172
**NUMBER OF EPISODES**

*"Typical Sontaran attitude. Stop, Linx..."*
**FIRST WORDS**

*"It's the end. But the moment has been prepared for."*
**LAST WORDS**

**SONIC SCREWDRIVER MARK III**

**APPEARANCES IN OTHER STORIES**

EARTHSHOCK • MAWDRYN UNDEAD
**THE FIVE DOCTORS • RESURRECTION OF THE DALEKS**
THE NEXT DOCTOR • THE ELEVENTH HOUR
**THE LODGER • NIGHTMARE IN SILVER**
THE ALMOST PEOPLE (VOICE)
**THE NAME OF THE DOCTOR • THE DAY OF THE DOCTOR**
THE WITCH'S FAMILIAR

---

**Top-right chart**

viewing figures (in millions) · Doctor Who Magazine poll (out of 241)
Some AI figures weren't collected for Season 15

- Horror of Fang Rock Part 1 — 6.8, 7.1
- Horror of Fang Rock Part 2 — 9.8
- Horror of Fang Rock Part 3 — 9.9
- Horror of Fang Rock Part 4 — 8.6
- The Invisible Enemy Part 1
- The Invisible Enemy Part 2 — 7.3, 7.5
- The Invisible Enemy Part 3 — 8.3
- The Invisible Enemy Part 4
- Image of the Fendahl Part 1
- Image of the Fendahl Part 2 — 7.5
- Image of the Fendahl Part 3 — 7.9
- Image of the Fendahl Part 4 — 7.1
- The Sun Makers Part 1 — 8.5
- The Sun Makers Part 2 — 9.5
- The Sun Makers Part 3 — 8.9
- The Sun Makers Part 4 — 8.9, 8.4
- Underworld Part 1 — 9.1
- Underworld Part 2 — 9.1
- Underworld Part 3 — 8.9
- Underworld Part 4 — 11.2, 11.7
- The Invasion of Time Part 1
- The Invasion of Time Part 2 — 11.4
- The Invasion of Time Part 3 — 9.5
- The Invasion of Time Part 4 — 10.9, 10.3
- The Invasion of Time Part 5
- The Invasion of Time Part 6 — 9.8

---

**Bottom-left chart**

viewing figures (in millions) · Doctor Who Magazine poll (out of 241)
Some AI figures weren't collected for Season 14

- The Masque of Mandragora Part 1 — 8.3
- The Masque of Mandragora Part 2 — 9.8, 9.2
- The Masque of Mandragora Part 3 — 10.6
- The Masque of Mandragora Part 4
- The Hand of Fear Part 1 — 10.5, 10.2, 11.1
- The Hand of Fear Part 2
- The Hand of Fear Part 3
- The Hand of Fear Part 4 — 12
- The Deadly Assassin Part 1 — 11.8
- The Deadly Assassin Part 2 — 12.1, 13
- The Deadly Assassin Part 3 — 13.1
- The Deadly Assassin Part 4 — 11.8
- The Face of Evil Part 1 — 10.7, 11.1
- The Face of Evil Part 2 — 11.3
- The Face of Evil Part 3 — 11.7
- The Face of Evil Part 4
- The Robots of Death Part 1 — 12.8
- The Robots of Death Part 2 — 12.4, 13.1
- The Robots of Death Part 3 — 12.6
- The Robots of Death Part 4
- The Talons of Weng-Chiang Part 1 — 11.3, 9.8
- The Talons of Weng-Chiang Part 2
- The Talons of Weng-Chiang Part 3 — 10.2, 11.4
- The Talons of Weng-Chiang Part 4
- The Talons of Weng-Chiang Part 5 — 10.1
- The Talons of Weng-Chiang Part 6 — 9.3

# THE FOURTH DOCTOR
## VITAL STATISTICS

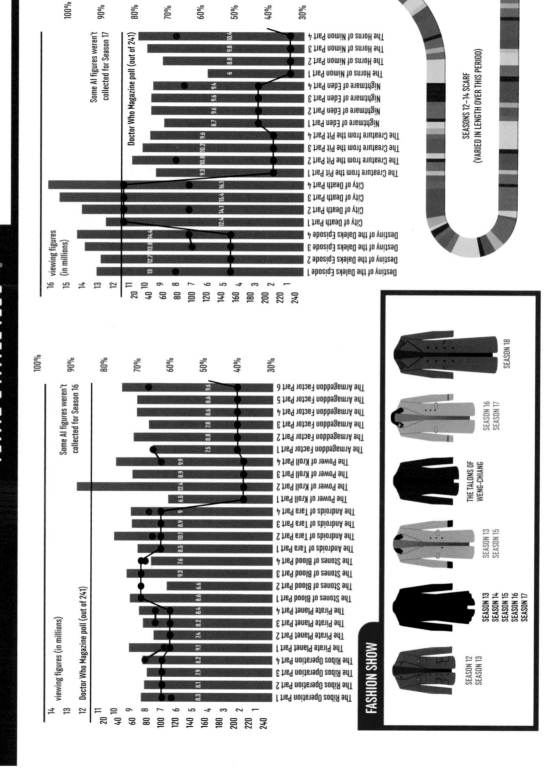

### (Top chart)

viewing figures (in millions)

Doctor Who Magazine poll (out of 24)

Some AI figures weren't collected for Season 17

- Destiny of the Daleks Episode 1 — 13
- Destiny of the Daleks Episode 2
- Destiny of the Daleks Episode 3 — 12.7, 13.8, 14.4
- Destiny of the Daleks Episode 4
- City of Death Part 1 — 12.4, 14.1, 15.4, 16.1
- City of Death Part 2
- City of Death Part 3
- City of Death Part 4
- The Creature from the Pit Part 1 — 9.3, 10.8, 10.2, 9.6
- The Creature from the Pit Part 2
- The Creature from the Pit Part 3
- The Creature from the Pit Part 4
- Nightmare of Eden Part 1 — 8.7, 9.6, 9.6, 9.4
- Nightmare of Eden Part 2
- Nightmare of Eden Part 3
- Nightmare of Eden Part 4
- The Horns of Nimon Part 1 — 6, 8.8, 8.9, 10.4
- The Horns of Nimon Part 2
- The Horns of Nimon Part 3
- The Horns of Nimon Part 4

### (Bottom chart)

viewing figures (in millions)

Doctor Who Magazine poll (out of 24)

Some AI figures weren't collected for Season 16

- The Ribos Operation Part 1 — 8.3
- The Ribos Operation Part 2 — 8.1
- The Ribos Operation Part 3 — 7.9
- The Ribos Operation Part 4 — 8.2
- The Pirate Planet Part 1 — 9.1
- The Pirate Planet Part 2 — 7.4
- The Pirate Planet Part 3 — 8.2
- The Pirate Planet Part 4 — 8.4
- The Stones of Blood Part 1 — 8.6
- The Stones of Blood Part 2 — 6.6
- The Stones of Blood Part 3 — 7.6
- The Stones of Blood Part 4 — 9.3
- The Androids of Tara Part 1 — 9
- The Androids of Tara Part 2 — 8.9, 10.1
- The Androids of Tara Part 3 — 8.5
- The Androids of Tara Part 4 — 9, 8.9, 9
- The Power of Kroll Part 1 — 6.5
- The Power of Kroll Part 2 — 12.4, 8.9
- The Power of Kroll Part 3 — 7.5
- The Power of Kroll Part 4 — 9.9
- The Armageddon Factor Part 1
- The Armageddon Factor Part 2 — 8.8, 7.8
- The Armageddon Factor Part 3 — 8.6
- The Armageddon Factor Part 4 — 8.6
- The Armageddon Factor Part 5 — 9.6
- The Armageddon Factor Part 6

### FASHION SHOW

- SEASON 12 / SEASON 13
- SEASON 13 / SEASON 14 / SEASON 15 / SEASON 16 / SEASON 17
- SEASON 13 / SEASON 15
- THE TALONS OF WENG-CHIANG
- SEASON 16 / SEASON 17
- SEASON 18

SEASONS 12–14 SCARF
(VARIED IN LENGTH OVER THIS PERIOD)

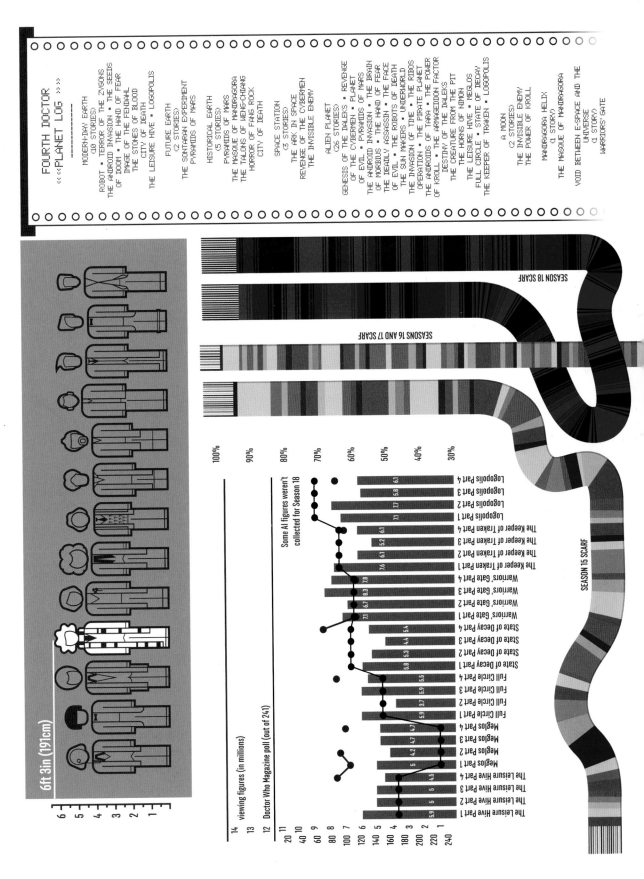

# FOURTH DOCTOR
## ::: PLANET LOG :::

**MODERN-DAY EARTH**
(10 STORIES)
ROBOT ▪ TERROR OF THE ZYGONS
THE ANDROID INVASION ▪ THE SEEDS
OF DOOM ▪ THE HAND OF FEAR
IMAGE OF THE FENDAHL
THE STONES OF BLOOD
CITY OF DEATH
THE LEISURE HIVE ▪ LOGOPOLIS

**FUTURE EARTH**
(2 STORIES)
THE SONTARAN EXPERIMENT
PYRAMIDS OF MARS

**HISTORICAL EARTH**
(5 STORIES)
PYRAMIDS OF MARS
THE MASQUE OF MANDRAGORA
THE TALONS OF WENG-CHIANG
HORROR OF FANG ROCK
CITY OF DEATH

**SPACE STATION**
(3 STORIES)
THE ARK IN SPACE
REVENGE OF THE CYBERMEN
THE INVISIBLE ENEMY

**ALIEN PLANET**
(26 STORIES)
GENESIS OF THE DALEKS ▪ REVENGE
OF THE CYBERMEN ▪ PLANET
OF EVIL ▪ PYRAMIDS OF MARS
THE ANDROID INVASION ▪ THE BRAIN
OF MORBIUS ▪ THE HAND OF FEAR
THE DEADLY ASSASSIN ▪ THE FACE
OF EVIL ▪ THE ROBOTS OF DEATH
THE SUN MAKERS ▪ UNDERWORLD
THE INVASION OF TIME ▪ THE RIBOS
OPERATION ▪ THE PIRATE PLANET
THE ANDROIDS OF TARA ▪ THE POWER
OF KROLL ▪ THE ARMAGEDDON FACTOR
DESTINY OF THE DALEKS
THE CREATURE FROM THE PIT
THE HORNS OF NIMON
THE LEISURE HIVE ▪ MEGLOS
FULL CIRCLE ▪ STATE OF DECAY
THE KEEPER OF TRAKEN ▪ LOGOPOLIS

**A MOON**
(2 STORIES)
THE INVISIBLE ENEMY
THE POWER OF KROLL

**MANDRAGORA HELIX**
(1 STORY)
THE MASQUE OF MANDRAGORA

**VOID BETWEEN E-SPACE AND THE UNIVERSE**
(1 STORY)
WARRIORS' GATE

6ft 3in (191cm)

6
5
4
3
2
1

SEASON 18 SCARF

SEASONS 16 AND 17 SCARF

SEASON 15 SCARF

100%
90%
80%
70%
60%
50%
40%
30%

14  viewing figures (in millions)

13  Some AI figures weren't collected for Season 18

12  Doctor Who Magazine poll (out of 241)

20   11
40   10
60   9
80   8
100  7
120  6
140  5
160  4
180  3
200  2
220  1
240

The Leisure Hive Part 1    5.9
The Leisure Hive Part 2    5
The Leisure Hive Part 3    5
The Leisure Hive Part 4    4.5
Meglos Part 1
Meglos Part 2    5
Meglos Part 3    4.2    4.7
Meglos Part 4
Full Circle Part 1    5.9    3.7
Full Circle Part 2    5.9
Full Circle Part 3    5.5
Full Circle Part 4    5.4
State of Decay Part 1    5.8
State of Decay Part 2    5.3
State of Decay Part 3    4.4
State of Decay Part 4    5.4
Warriors' Gate Part 1    7.1
Warriors' Gate Part 2    6.7
Warriors' Gate Part 3    8.3
Warriors' Gate Part 4    7.8
The Keeper of Traken Part 1    7.6
The Keeper of Traken Part 2    6.1
The Keeper of Traken Part 3    5.2
The Keeper of Traken Part 4    6.1
Logopolis Part 1    7.1
Logopolis Part 2    7.7
Logopolis Part 3    5.8
Logopolis Part 4    6.1

**c. 29 June 1979**
Doctor Who script editor Douglas Adams begins work on Sunburst, a six-part Doctor Who story intended to be broadcast in January and February 1980.

**15 October 1979**
Filming begins in Cambridge on the story, now called Shada, with Tom Baker as the Fourth Doctor and Lalla Ward as companion Romana.

**19 November 1979**
The cast and crew are locked out of the recording studio due to industrial action at the BBC.

**25 November 1983**
The Five Doctors is broadcast – the first time any material from Shada is aired.

**September 1983**
"The story you never saw!" declares the cover of issue 81 of Doctor Who Monthly (cover dated October), with photographs, interviews and features telling the story of Shada's cancellation.

1984

1985

1979

1986

**June 1987**
Publication of Douglas Adams's novel Dirk Gently's Holistic Detective Agency, which reuses some material from Shada (but not the Doctor and Romana).

1987

**September 1987**
Release of an abridged audio version of Dirk Gently's Holistic Detective Agency, read by Douglas Adams.

# Shada, Shada, Shada!

The repeatedly completed uncompleted adventure...

1989

1988

**5 January 1992**
Characters from Dirk Gently's Holistic Detective Agency appear in an episode of ITV arts documentary series The South Bank Show devoted to Douglas Adams.

1991

1992

1990

**7 January 2013**
The 1992 version of Shada is released on DVD as part of The Legacy Collection, together with the 2003 animated version.

**15 March 2012**
BBC Books publish a new novelisation of Shada by Gareth Roberts, and AudioGO release an unabridged audiobook version read by Lalla Ward.

**12 March 2012**
An episode of the TV series Dirk Gently sees him return to his old university – St Cedd's College, Cambridge, as featured in Shada.

2013

2012

2014

**2 May 2015**
The first in a Dirk Gently comic-strip series is released in the US, written by Chris Ryall, with pencils and inks by Tony Akins and colour by Leonard O'Grady.

2015

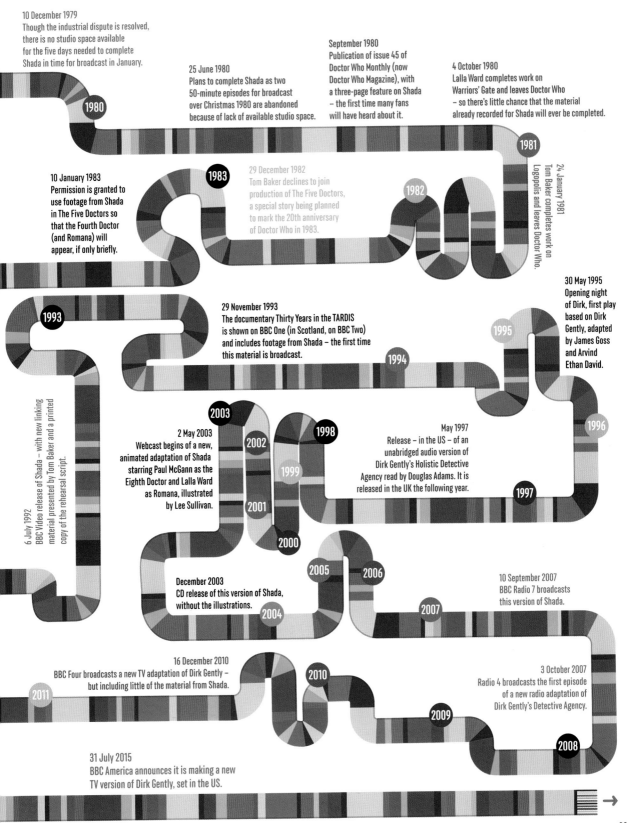

**10 December 1979**
Though the industrial dispute is resolved, there is no studio space available for the five days needed to complete Shada in time for broadcast in January.

**25 June 1980**
Plans to complete Shada as two 50-minute episodes for broadcast over Christmas 1980 are abandoned because of lack of available studio space.

**September 1980**
Publication of issue 45 of Doctor Who Monthly (now Doctor Who Magazine), with a three-page feature on Shada – the first time many fans will have heard about it.

**4 October 1980**
Lalla Ward completes work on Warriors' Gate and leaves Doctor Who – so there's little chance that the material already recorded for Shada will ever be completed.

**1980**

**1981**

**24 January 1981**
Tom Baker completes work on Logopolis and leaves Doctor Who.

**10 January 1983**
Permission is granted to use footage from Shada in The Five Doctors so that the Fourth Doctor (and Romana) will appear, if only briefly.

**1983**

**29 December 1982**
Tom Baker declines to join production of The Five Doctors, a special story being planned to mark the 20th anniversary of Doctor Who in 1983.

**1982**

**30 May 1995**
Opening night of Dirk, first play based on Dirk Gently, adapted by James Goss and Arvind Ethan David.

**1993**

**29 November 1993**
The documentary Thirty Years in the TARDIS is shown on BBC One (in Scotland, on BBC Two) and includes footage from Shada – the first time this material is broadcast.

**1994**

**1995**

**6 July 1992**
BBC Video release of Shada – with new linking material presented by Tom Baker and a printed copy of the rehearsal script.

**2003**

**2 May 2003**
Webcast begins of a new, animated adaptation of Shada starring Paul McGann as the Eighth Doctor and Lalla Ward as Romana, illustrated by Lee Sullivan.

**2002**

**1998**

**1999**

**2001**

**2000**

**May 1997**
Release – in the US – of an unabridged audio version of Dirk Gently's Holistic Detective Agency read by Douglas Adams. It is released in the UK the following year.

**1996**

**1997**

**December 2003**
CD release of this version of Shada, without the illustrations.

**2005**

**2006**

**2004**

**10 September 2007**
BBC Radio 7 broadcasts this version of Shada.

**2007**

**16 December 2010**
BBC Four broadcasts a new TV adaptation of Dirk Gently – but including little of the material from Shada.

**2010**

**2011**

**3 October 2007**
Radio 4 broadcasts the first episode of a new radio adaptation of Dirk Gently's Detective Agency.

**2009**

**2008**

**31 July 2015**
BBC America announces it is making a new TV version of Dirk Gently, set in the US.

Susan • Barbara • Ian • Vicki • Steven • Katarina • Sara • Dodo (short for Dorothea) • Polly • Ben (short for Benjamin) • Jamie • Victoria • Zoe • Liz (short for Elizabeth) • Jo (short for Josephine) • Sarah Jane • Harry • Leela • K-9 • Romana I • Romana II • Adric • Nyssa • Tegan • Turlough • Kamelion • Peri (short for Perpugilliam) • Mel (short for Melanie) • Ace (nickname of Dorothy) • Grace • Rose • Adam • Jack • Mickey • Martha • Donna • Amy (short for Amelia) • Rory • Clara •

Alistair (the Brigadier) • Chang Lee • Jackie (short for Jacqueline) • Astrid • Wilf (short for Wilfrid) • Christina • Adelaide • River (also known as Mels or Melanie) • Kate •

# Say "Aaaa"

The frequency of the letter A in the names of companions.

The Doctor and Sarah protect themselves from an exploding nuclear power plant by hiding behind a jeep
**The Hand of Fear**

The Doctor hangs from a guardrail. For no reason
**Dragonfire**

Romana regenerates... multiple times
**Destiny of the Daleks**

The burping wheelie bin
**Rose**

Sarah trips over a mound of earth
**The Five Doctors**

The Doctor disguises himself as a cleaning woman
**The Green Death**

The Doctor makes a 'time flow analogue' from a wine bottle, spoons, forks, corks, keyrings, tea leaves and a mug
**The Time Monster**

The TARDIS tows the Earth back to our solar system
**Journey's End**

Jo confuses Azal to death
**The Dæmons**

# DOCTOR WHO'S
# DAFTEST
## MOMENTS

The Queen thanks the Doctor for saving Buckingham Palace
Voyage of the Damned

Solow karate kicks the Myrka
Warriors of the Deep

The Doctor plans to flush the Master out of the TARDIS by materialising under the Thames
Logopolis

Peri is accosted by a tree
The Mark of the Rani

Hobson and Benoit stop the moonbase from depressurising by plugging the leak with a coffee tray
The Moonbase

The Rani disguises herself as Mel
Time and the Rani

# Puffball

The names of Fourth Doctor stories that almost were.

The Android Invasion
The Androids of Tara
The Ark in Space Part 1
The Ark in Space Part 2
The Ark in Space Part 3
The Ark in Space Part 4
The Armageddon Factor
City of Death
The Deadly Assassin
The Face of Evil
Full Circle
Genesis of the Daleks
The Hand of Fear
Horror of Fang Rock
The Invasion of Time
The Invisible Enemy
The Leisure Hive
The Masque of Mandragora
Meglos
The Pirate Planet
The Power of Kroll
Revenge of the Cybermen
Revenge of the Cybermen Part 1
Revenge of the Cybermen Part 2
Revenge of the Cybermen Part 3
Revenge of the Cybermen Part 4
The Ribos Operation
The Robots of Death
The Sontaran Experiment
State of Decay
The Stones of Blood
The Talons of Weng-Chiang
Terror of the Zygons
Warriors' Gate

● Broadcast Title for Story
● Working Title for Story

\* We've not included small variations to final titles,
   such as adding or removing the word "the".

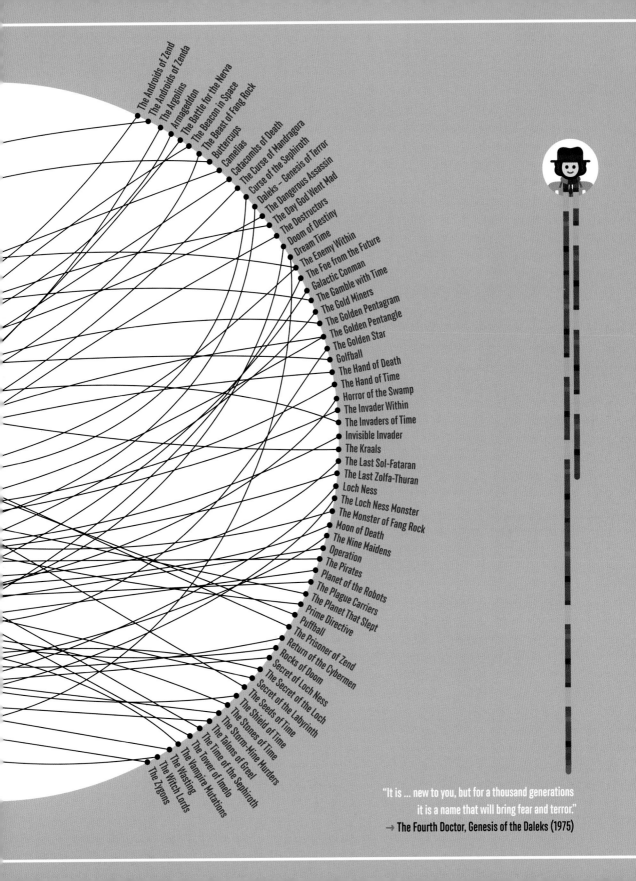

The Androids of Zend
The Androids of Zenda
The Argolins
Armageddon
The Battle for the Nerva
The Beacon in Space
The Beast of Fang Rock
Buttercups
Camelias
Catacombs of Death
The Curse of Mandragora
Curse of the Sephiroth
Daleks – Genesis of Terror
The Dangerous Assassin
The Day God Went Mad
The Destructors
Doom of Destiny
Dream Time
The Enemy Within
The Foe from the Future
Galactic Conman
The Gamble with Time
The Gold Miners
The Golden Pentagram
The Golden Pentangle
The Golden Star
Golfball
The Hand of Death
The Hand of Time
Horror of the Swamp
The Invader Within
The Invaders of Time
Invisible Invader
The Kraals
The Last Sol-Fataran
The Last Zolfa-Thuran
Loch Ness
The Loch Ness Monster
The Monster of Fang Rock
Moon of Death
The Nine Maidens
Operation
The Pirates
Planet of the Robots
The Plague Carriers
The Planet That Slept
Prime Directive
Puffball
The Prisoner of Zend
Return of the Cybermen
Rocks of Doom
Secret of Loch Ness
The Secret of the Loch
Secret of the Labyrinth
The Seeds of Time
The Shield of Time
The Stones of Time
The Storm-Mine Murders
The Talons of the Sephiroth
The Time of Imelo
The Tower of Imelo
The Vampire Mutations
The Wasting
The Witch Lords
The Zygons

"It is ... new to you, but for a thousand generations
it is a name that will bring fear and terror."
→ The Fourth Doctor, Genesis of the Daleks (1975)

# A Tune!

The Doctor's record of performing music.

**HARP**

THE FIVE DOCTORS (1983)

**LYRE**

THE ROMANS (1965)

ATTACK OF THE CYBERMEN (1985)

THE LAZARUS EXPERIMENT (2007)

**ORGAN**

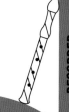

**RECORDER**

THE POWER OF THE DALEKS (1966)

THE HIGHLANDERS (1966–7)

THE UNDERWATER MENACE (1967)

THE MACRA TERROR (1967)

THE EVIL OF THE DALEKS (1967)

THE ABOMINABLE SNOWMEN (1967)

THE THREE DOCTORS (1972–3)

**GUITAR**

THE MAGICIAN'S APPRENTICE (2015)

BEFORE THE FLOOD (2015)

THE WOMAN WHO LIVED (2015)

THE ZYGON INVASION (2015)

HELL BENT (2015)

## FAMOUS MUSICIANS THE DOCTOR HAS APPARENTLY MET

BENJAMIN FRANKLIN* (1706–1790)
→ Smith and Jones (2007)

LUDWIG VAN BEETHOVEN (1770–1827)
→ The Lazarus Experiment (2007),
Before the Flood (2015)

FRANZ SCHUBERT (1797–1828)
→ Dinosaurs on a Spaceship (2012)

WS GILBERT (1836–1911) AND
ARTHUR SULLIVAN (1842–1900)
→ The Edge of Destruction (1964)

GIACOMO PUCCINI (1858–1924)
→ Doctor Who (1996)

JANIS JOPLIN (1943–70)
→ Gridlock (2007)

FRANK SINATRA (1915–98)
→ A Christmas Carol (2010)

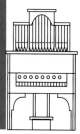

*(inventor of the glass armonica)

**REED PIPE**

THE POWER OF KROLL (1978–9)

**WHISTLES**

THE ENEMY OF THE WORLD (1968)

THE FACE OF EVIL (1977)

THE TALONS OF WENG-CHIANG (1977)

THE INVASION OF TIME (1978)

DESTINY OF THE DALEKS (1979)

**SINGS**

"Dramatic recitations, singing, tap-dancing. I can play the trumpet voluntary in a bowl of live goldfish."
→ The Fourth Doctor, The Talons of Weng-Chiang (1977)

THE CHASE (1965)

SPEARHEAD FROM SPACE (1970)     DOCTOR WHO AND THE SILURIANS (1970)

INFERNO (1970)     TERROR OF THE AUTONS (1971)     THE CURSE OF PELADON (1972)

DEATH TO THE DALEKS (1974)     THE MONSTER OF PELADON (1974)

THE POWER OF KROLL (1978–9)

BLACK ORCHID (1982)

THE FIVE DOCTORS (1983)

THE TWO DOCTORS (1985)     THE TRIAL OF A TIME LORD (1986)

THE HAPPINESS PATROL (1988)

THE GIRL IN THE FIREPLACE (2006)

THE LODGER (2010)

**SPOONS**

TIME AND THE RANI (1987)

THE HAPPINESS PATROL (1988)

THE GREATEST SHOW IN THE GALAXY (1988)

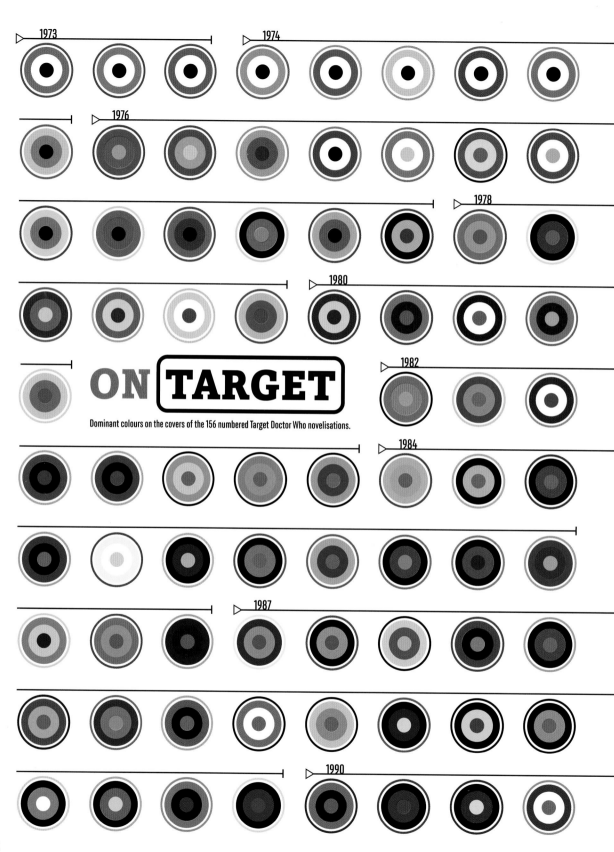

# ON TARGET

Dominant colours on the covers of the 156 numbered Target Doctor Who novelisations.

1973 1974 1976 1978 1980 1982 1984 1987 1990

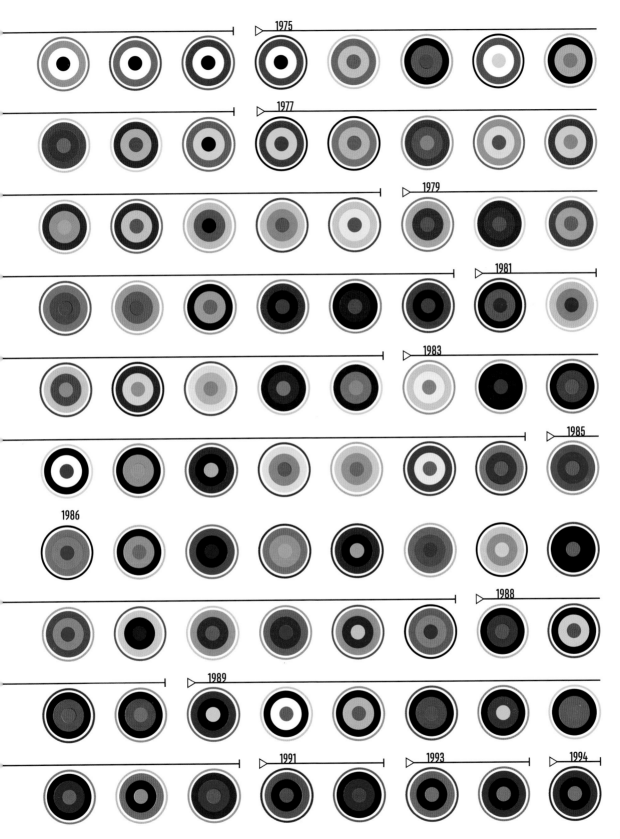

1975
1977
1979
1981
1983
1985
1986
1988
1989
1991
1993
1994

91

# SATURDAYNESS

Days of the week on which the 826 episodes of Doctor Who were first broadcast.

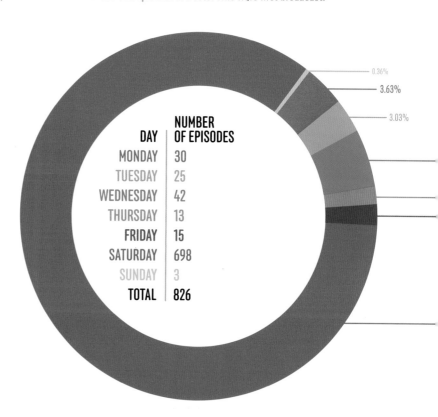

0.36%

3.63%

3.03%

| DAY | NUMBER OF EPISODES |
|---|---|
| MONDAY | 30 |
| TUESDAY | 25 |
| WEDNESDAY | 42 |
| THURSDAY | 13 |
| FRIDAY | 15 |
| SATURDAY | 698 |
| SUNDAY | 3 |
| TOTAL | 826 |

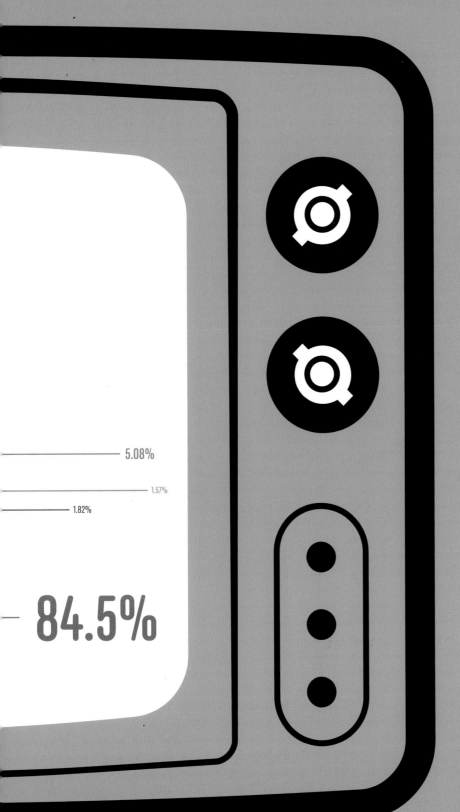

5.08%

1.57%

1.82%

# 84.5%

"There's loads of boring stuff like Sundays and Tuesdays and Thursday afternoons. But now and then there are Saturdays: big temporal tipping points when anything's possible. The TARDIS can't resist them, like a moth to a flame."
→ The Eleventh Doctor, The Impossible Astronaut (2011)

# CHAPTER ONE
## THE DOCTOR

# CHAPTER TWO
## MONSTERS AND ENEMIES

# CHAPTER SIX
## TOTAL DESTRUCTION

# CHAPTER THREE
## EARTH

# CHAPTER TEN
## TARDIS

# CHAPTER FIVE
## FRIENDS AND COMPANIONS

# THE FIFTH DOCTOR
## VITAL STATISTICS

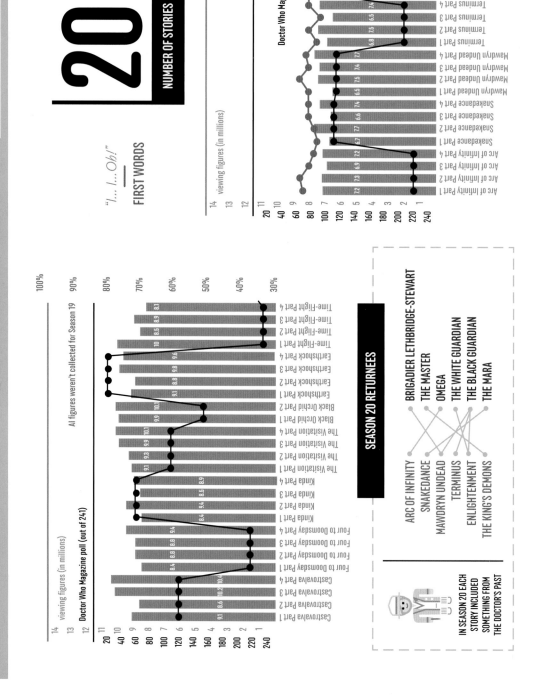

**FIRST WORDS**
"I... I... Oh!"

**LAST WORDS**
"Adric?"

**NUMBER OF STORIES**
# 20

### SEASON 20 RETURNEES

ARC OF INFINITY — BRIGADIER LETHBRIDGE-STEWART
SNAKEDANCE — THE MASTER
MAWDRYN UNDEAD — OMEGA
TERMINUS — THE WHITE GUARDIAN
ENLIGHTENMENT — THE BLACK GUARDIAN
THE KING'S DEMONS — THE MARA

IN SEASON 20 EACH STORY INCLUDED SOMETHING FROM THE DOCTOR'S PAST

---

viewing figures (in millions)
Doctor Who Magazine poll (out of 241)

**Castrovalva Part 1** 9.1
**Castrovalva Part 2** 8.6
**Castrovalva Part 3** 10.2
**Castrovalva Part 4** 10.4
**Four to Doomsday Part 1** 8.4
**Four to Doomsday Part 2** 8.8
**Four to Doomsday Part 3** 8.8
**Four to Doomsday Part 4** 9.4
**Kinda Part 1** 8.4
**Kinda Part 2** 9.4
**Kinda Part 3** 8.5
**Kinda Part 4** 8.9
**The Visitation Part 1** 9.1
**The Visitation Part 2** 9.3
**The Visitation Part 3** 9.9
**The Visitation Part 4** 10.1
**Black Orchid Part 1** 9.9
**Black Orchid Part 2** 10.1
**Earthshock Part 1** 9.1
**Earthshock Part 2** 8.8
**Earthshock Part 3** 9.8
**Earthshock Part 4** 9.6
**Time-Flight Part 1** 10
**Time-Flight Part 2** 8.5
**Time-Flight Part 3** 8.9
**Time-Flight Part 4** 8.1

All figures weren't collected for Season 19

---

viewing figures (in millions)
Doctor Who Magazine poll (out of 241)

**Arc of Infinity Part 1** 7.2
**Arc of Infinity Part 2** 7.3
**Arc of Infinity Part 3** 6.9
**Arc of Infinity Part 4** 7.2
**Snakedance Part 1** 6.7
**Snakedance Part 2** 7.7
**Snakedance Part 3** 6.6
**Snakedance Part 4** 7.4
**Mawdryn Undead Part 1** 6.5
**Mawdryn Undead Part 2** 7.5
**Mawdryn Undead Part 3** 7.4
**Mawdryn Undead Part 4** 7.7
**Terminus Part 1** 6.8
**Terminus Part 2** 7.5
**Terminus Part 3** 6.5
**Terminus Part 4** 7.4
**Enlightenment Part 1** 6.6
**Enlightenment Part 2** 7.2
**Enlightenment Part 3** 6.2
**Enlightenment Part 4** 7.3
**The King's Demons Part 1** 5.8
**The King's Demons Part 2** 7.2
**The Five Doctors** 7.7

All figures

# 69

## NUMBER OF EPISODES

SONIC SCREWDRIVER MARK III

6ft 0.23in (184cm)

6
5
4
3
2
1

## FASHION SHOW

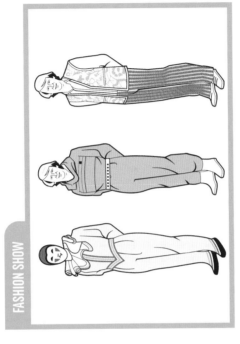

## FIVE'S FRIENDS

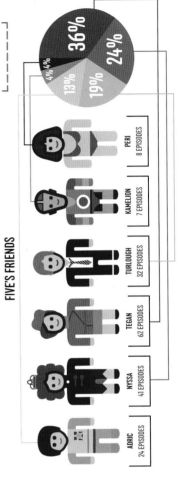

36%

24%

19%

13%

4%  4%

| ADRIC 24 EPISODES | NYSSA 41 EPISODES | TEGAN 62 EPISODES | TURLOUGH 32 EPISODES | KAMELION 7 EPISODES | PERI 8 EPISODES |

### viewing figures (in millions)
### Doctor Who Magazine poll (out of 241)

Some AI figures weren't collected for Season 21

100%
90%
80%
70%
60%
50%
40%
30%

14
13
12
11
10
9
8
7
6
5
4
3
2
1

20
40
60
80
100
120
140
160
180
200
220
240

Warriors of the Deep Part 1
Warriors of the Deep Part 2 — 7.6
Warriors of the Deep Part 3 — 7.5
Warriors of the Deep Part 4 — 7.3
The Awakening Part 1 — 6.6
The Awakening Part 2 — 7.9
Frontios Part 1 — 8
Frontios Part 2 — 5.8
Frontios Part 3 — 7.8
Frontios Part 4 — 5.6
Resurrection of the Daleks Part 1 — 7.3
Resurrection of the Daleks Part 2 — 8
Planet of Fire Part 1 — 7.4
Planet of Fire Part 2 — 6.1
Planet of Fire Part 3 — 7.4
Planet of Fire Part 4 — 7
The Caves of Androzani Part 1 — 6.9
The Caves of Androzani Part 2 — 6.6
The Caves of Androzani Part 3 — 7.8
The Caves of Androzani Part 4 — 7.8

# COMPANION COUNT

The most frequently appearing of the Doctor's friends.

Ian Chesterton    Barbara Wright    Jamie McCrimmon    Alistair Gordon Lethbridge-Stewart

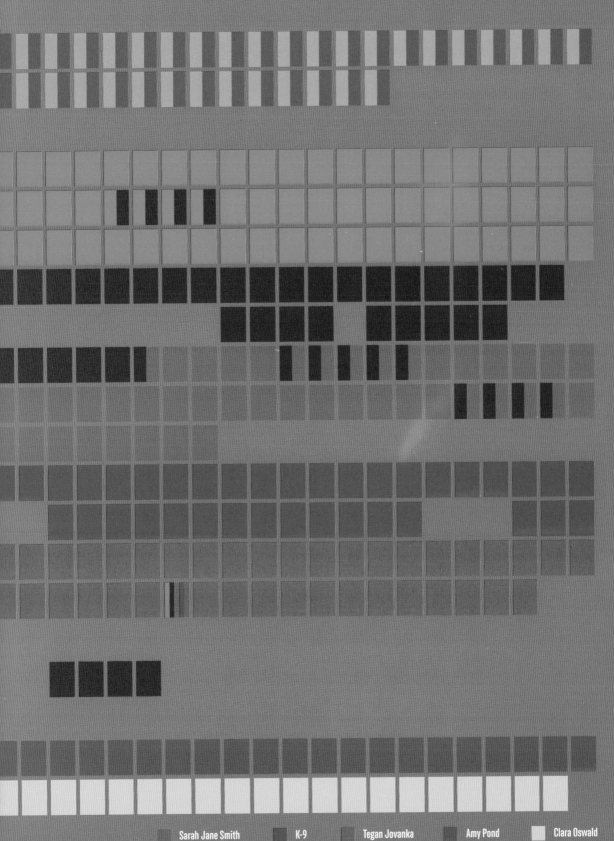

Sarah Jane Smith   K-9   Tegan Jovanka   Amy Pond   Clara Oswald   99

# COMPANION EXITS

So long, farewell, auf wiedersehen, goodbye.

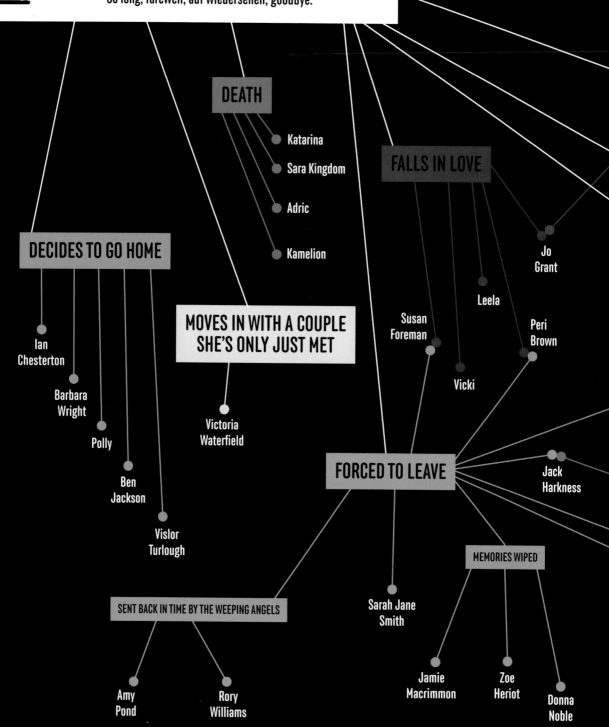

**DEATH**
- Katarina
- Sara Kingdom
- Adric
- Kamelion

**FALLS IN LOVE**
- Jo Grant
- Leela
- Peri Brown
- Susan Foreman
- Vicki

**DECIDES TO GO HOME**
- Ian Chesterton
- Barbara Wright
- Polly
- Ben Jackson
- Vislor Turlough

**MOVES IN WITH A COUPLE SHE'S ONLY JUST MET**
- Victoria Waterfield

**FORCED TO LEAVE**
- Jack Harkness
- Sarah Jane Smith

**MEMORIES WIPED**
- Jamie Macrimmon
- Zoe Heriot
- Donna Noble

**SENT BACK IN TIME BY THE WEEPING ANGELS**
- Amy Pond
- Rory Williams

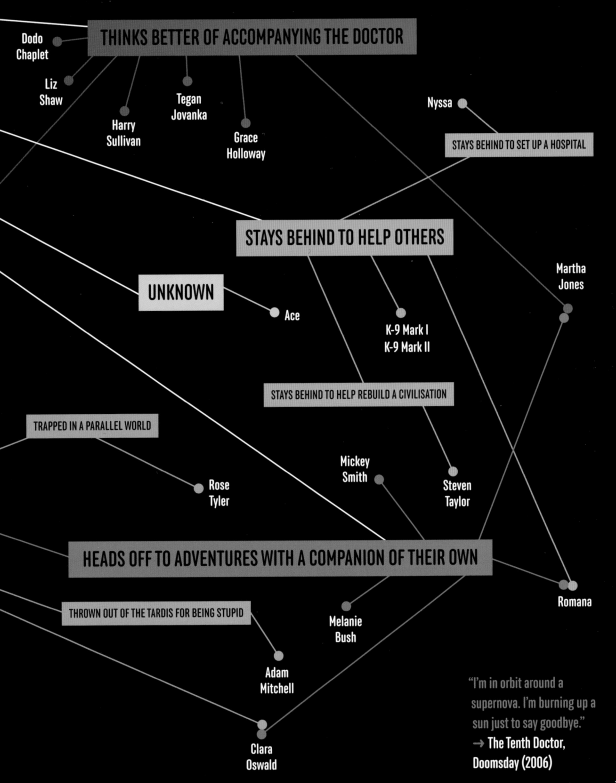

THINKS BETTER OF ACCOMPANYING THE DOCTOR

Dodo Chaplet

Liz Shaw

Harry Sullivan

Tegan Jovanka

Grace Holloway

Nyssa

STAYS BEHIND TO SET UP A HOSPITAL

STAYS BEHIND TO HELP OTHERS

Martha Jones

UNKNOWN

Ace

K-9 Mark I
K-9 Mark II

STAYS BEHIND TO HELP REBUILD A CIVILISATION

TRAPPED IN A PARALLEL WORLD

Rose Tyler

Mickey Smith

Steven Taylor

HEADS OFF TO ADVENTURES WITH A COMPANION OF THEIR OWN

Romana

THROWN OUT OF THE TARDIS FOR BEING STUPID

Melanie Bush

Adam Mitchell

"I'm in orbit around a supernova. I'm burning up a sun just to say goodbye."
→ The Tenth Doctor, Doomsday (2006)

Clara Oswald

# Companions' JOBS

What their jobs are when they meet the Doctor and what they end up doing after...

Susan Foreman
Barbara Wright
Ian Chesterton
Vicki
Steven Taylor
Katarina
Sara Kingdom
Dodo Chaplet
Polly
Ben Jackson
Jamie McCrimmon
Victoria Waterfield
Zoe Heriot
Liz Shaw
Jo Grant
Sarah Jane Smith
Harry Sullivan
Leela
K-9
Romana
Adric
Nyssa
Tegan Jovanka
Turlough
Kamelion
Peri Brown
Mel Bush
Ace
Grace Holloway
Rose Tyler
Adam Mitchell
Jack Harkness
Mickey Smith
Donna Noble
Martha Jones
Amelia Pond
Rory Williams
Clara Oswald

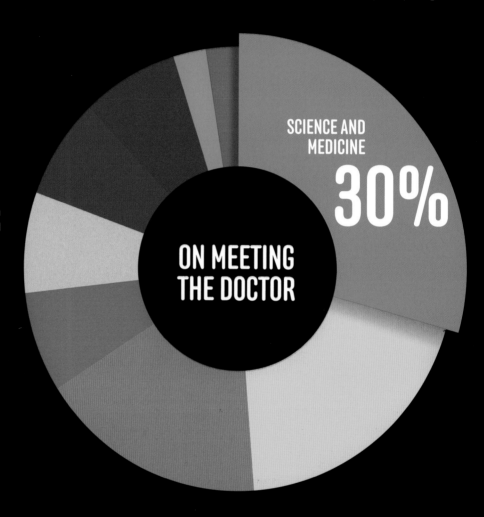

ON MEETING THE DOCTOR

SCIENCE AND MEDICINE

30%

SCIENCE AND MEDICINE **30%**
MILITARY/ SECURITY **18%**
SECRETARIAL/ WRITING **7%**
TEACHING/ LOOKING AFTER CHILDREN **7%**
ADVENTURER **2%**

SERVICE INDUSTRIES (SHOP CLERK, STEWARDING, HANDMAIDENING) **20%**
UNKNOWN **7%**
STUDENT **7%**
SAILOR **2%**

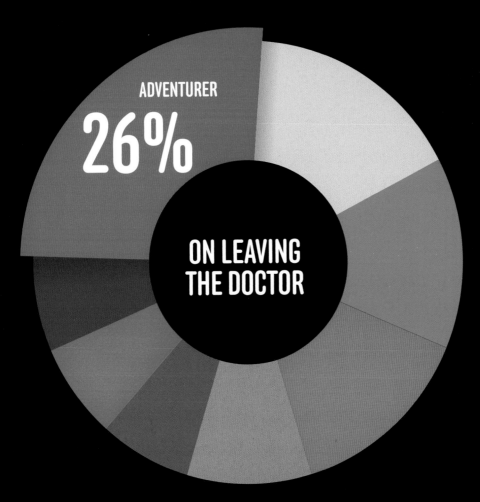

"Did I ever tell you, best temp in Chiswick?
Hundred words per minute!"
→ Donna Noble, Journey's End (2008)

ADVENTURER

26%

ON LEAVING
THE DOCTOR

ADVENTURER 26%
SCIENCE AND MEDICINE 14%
DEAD 9%
ROYALTY 7%

UNKNOWN 16%
MILITARY/ SECURITY 14%
SECRETARIAL/ WRITING 7%
TEACHING/ LOOKING AFTER CHILDREN 7%

Susan Foreman ●
Barbara Wright ●
Ian Chesterton ● ●
Vicki ●
Steven Taylor ●
Katarina ●
Sara Kingdom ●
Dodo Chaplet ●
Polly ●
Ben Jackson ●
Jamie McCrimmon ●
Victoria Waterfield ●
Zoe Heriot ●
Liz Shaw ●
Jo Grant ●
Sarah Jane Smith ● ●
Harry Sullivan ● ●
Leela ●
K-9 ●
Romana ●
Adric ●
Nyssa ●
Tegan Jovanka ●
Turlough ●
Kamelion ●
Peri Brown ●
Mel Bush ●
Ace ●
Grace Holloway ●
Rose Tyler ● ●
Adam Mitchell ●
Jack Harkness ● ●
Mickey Smith ● ●
Donna Noble ● ●
Martha Jones ● ●
Amelia Pond ●
Rory Williams ●
Clara Oswald

# Sarah Jane's
## Adventures

Since she met the Doctor,
Sarah never knows if she's
coming or going – or been.

**c. 2950**
Revenge of the Cybermen
The Doctor says in The Ark in Space
that the technology of
the Beacon means it was
built in the early 30th century.

**1951**
The Temptation of Sarah Jane Smith
Sarah is born in May 1951.

**c. 1250**
The Time Warrior
In The Sontaran Experiment
Sarah says Linx was
"destroyed in the 13th century."

**c. 1480**
The Masque of Mandragora
The Doctor says it's late 15th century,
not disputed by Giuliano who hears him.

**1978**
The Time Warrior
She says in the next story,
Invasion of the Dinosaurs, that she is 23. **

**1911**
Pyramids of Mars

| 1 | MAY 1951 SARAH JANE SMITH BORN | 7 | 1911 (AGAIN) |
|---|---|---|---|
| 2 | MEETS THE THIRD DOCTOR | 8 | LEAVES THE FOURTH DOCTOR |
| 3 | THE DOCTOR REGENERATES | 9 | MEETS THE TENTH DOCTOR |
| 4 | FURTHEST WE SEE SARAH TRAVEL INTO THE FUTURE | 10 | 18 AUGUST 1951 |
| 5 | 1911 | 11 | MEETS THE ELEVENTH DOCTOR |
| 6 | 1980 | 12 | 1889 |

**In Pyramids of Mars (1975), Sarah says she
is from 1980 but could be rounding up – or
by then has known the Doctor for some years.

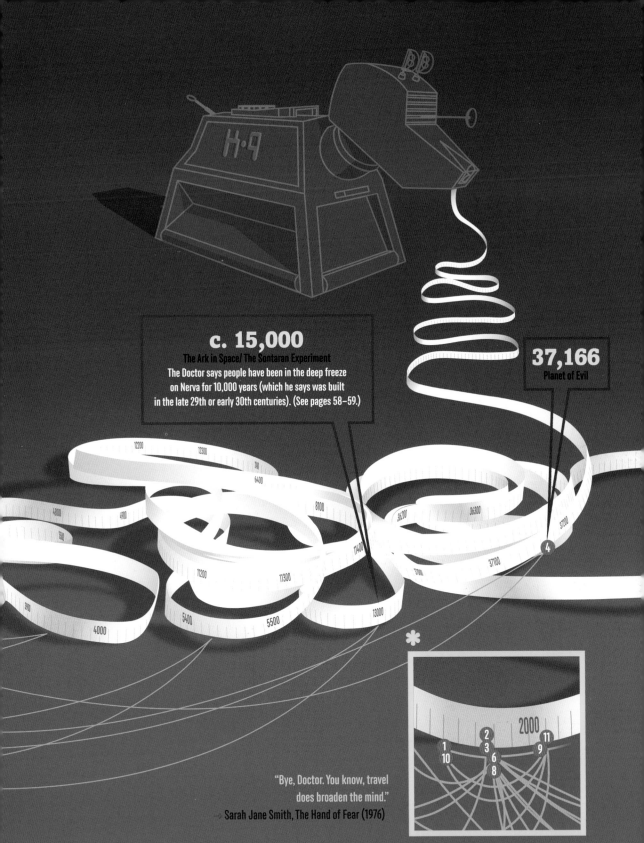

**c. 15,000**
The Ark in Space/ The Sontaran Experiment
The Doctor says people have been in the deep freeze
on Nerva for 10,000 years (which he says was built
in the late 29th or early 30th centuries). (See pages 58–59.)

**37,166**
Planet of Evil

"Bye, Doctor. You know, travel
does broaden the mind."
→ Sarah Jane Smith, The Hand of Fear (1976)

# DOUBLES

"I know where I got this face, and I know what it's for…"
→ The Twelfth Doctor, The Girl Who Died (2015)

## PEOPLES OF THE UNIVERSE

**UNNAMED WOMAN**
Soothsayer from Pompeii,
Roman Empire, seen AD 79

**LOBUS CAECILIUS**
Marble trader from Pompeii,
Roman Empire, seen AD 79

**JOANNA (OR JOAN)**
Queen of Sicily (and sister of Richard I of England), seen c. AD 1191

**THE ABBOT OF AMBOISE**
Abbot from Amboise, France, seen AD 1572

**ANN TALBOT**
From the vicinity of Cranleigh Halt, UK, seen AD 1925

**JOHN ANDREWS**
Naval lieutenant on the SS Bernice, seen AD 1926

**MORTON DILL**
Tourist from Alabama, USA, seen c. AD 1965

## DOCTORS AND COMPANIONS

**THE FIRST DOCTOR**

**BARBARA WRIGHT**

**STEVEN TAYLOR**

**SARA KINGDOM**

**THE SECOND DOCTOR**

**ALISTAIR LETHBRIDGE-STEWART
(THE BRIGADIER)**

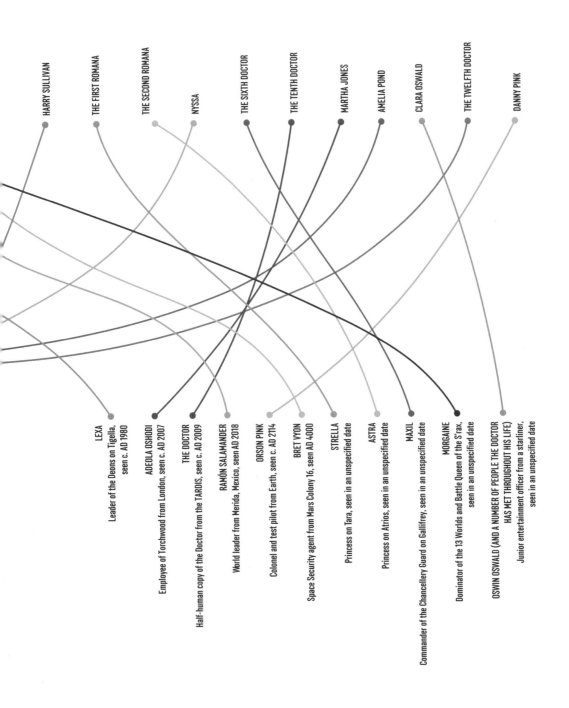

HARRY SULLIVAN

THE FIRST ROMANA

THE SECOND ROMANA

NYSSA

THE SIXTH DOCTOR

THE TENTH DOCTOR

MARTHA JONES

AMELIA POND

CLARA OSWALD

THE TWELFTH DOCTOR

DANNY PINK

LEXA
Leader of the Deons on Tigella,
seen c. AD 1980

ADEOLA OSHODI
Employee of Torchwood from London, seen c. AD 2007

THE DOCTOR
Half-human copy of the Doctor from the TARDIS, seen c. AD 2009

RAMÓN SALAMANDER
World leader from Merida, Mexico, seen AD 2018

ORSON PINK
Colonel and test pilot from Earth, seen c. AD 2114

BRET VYON
Space Security agent from Mars Colony 16, seen AD 4000

STRELLA
Princess on Tara, seen in an unspecified date

ASTRA
Princess on Atrios, seen in an unspecified date

MAXIL
Commander of the Chancellery Guard on Gallifrey, seen in an unspecified date

MORGAINE
Dominator of the 13 Worlds and Battle Queen of the S'rax,
seen in an unspecified date

OSWIN OSWALD (AND A NUMBER OF PEOPLE THE DOCTOR
HAS MET THROUGHOUT HIS LIFE)
Junior entertainment officer from a starliner,
seen in an unspecified date

**14.77%**
of episodes have women
villains or antagonists

# Roles
## FOR Women

Ratios of episodes with women companions,
and women villains or antagonists.

# 96.85%
of episodes have women companions

must ask him to credit me for that story idea

looks a bit like the Brigadier

dear Cleo!

ghastly old goat!

must say thanks for the coat

must get that laser spanner back off her!

Alexander the Great
Archimedes
Hans Christian Anderson
Marie Antoinette
Ludwig Van Beethoven
Napoleon Bonaparte
Isambard Kingdom Brunel
'Beau' Brummell

Cleopatra
Christopher Columbus
Marie Curie
Sir Francis Drake
Edward VII
Albert Einstein
Benjamin Franklin
Sigmund Freud
Gilbert and Sullivan

Hannibal

Henry VIII
Harry Houdini
Thomas Jefferson
Janis Joplin

David Lloyd George
Michelangelo
Mao Tse Tung

Horatio Nelson
Isaac Newton

Emmeline Pankhurst
Fred Perry
Pablo Picasso
Giacomo Puccini
Franz Schubert

Sir Walter Raleigh
William Tell

Leonardo da Vinci
James Watt

# The Name-dropping Doctor

note to self:
say no to a drink!

→ lovely fellow

Stories in which
I dropped their names

The Edge of Destruction
The Sensorites
The Romans
The Space Museum
Inferno
The Mind of Evil
Day of the Daleks
The Sea Devils
Planet of the Spiders
Robot
Revenge of the Cybermen
Pyramids of Mars
The Face of Evil
The Pirate Planet
The Stones of Blood

City of Death
Four to Doomsday
The Two Doctors

Doctor Who
Aliens of London
The Girl in the Fireplace
Smith and Jones
Gridlock
The Lazarus Experiment
Vincent and the Doctor
The Impossible Astronaut
Dinosaurs on a Spaceship
The Power of Three

"'Boney', I said, 'An army marches on its stomach.'"
→ The Third Doctor, Day of the Daleks

# Say My Name

Characters named in episode and story titles

1. Marco Polo (1964) 2. Mighty Kublai Khan (1964) 3. The Death of Doctor Who (1965) 4. The Meddling Monk (1965) 5. The Feast of Steven (1965) 6. The Celestial Toymaker (1966) 7. A Holiday for the Doctor (1966) 8. Johnny Ringo (1966) 9. Doctor Who and the Silurians (1970) 10. The Three Doctors (1972–3) 11. The Brain of Morbius (1976) 12. The Masque of Mandragora (1976) 13. The Power of Kroll (1978–9) 14. Meglos (1980) 15. Mawdryn Undead (1983) 16. The Five Doctors (1983) 17. The Mark of the Rani (1985) 18. The Two Doctors (1985) 19. Time and the Rani (1987) 20. Delta and the Bannermen (1987) 21. Ghost Light (1989) 22. The Curse of Fenric (1989) 23. Doctor Who (1996) 24. Rose (2005) 25. The Doctor Dances (2005) 26. Smith and Jones (2007) 27. The Shakespeare Code (2007) 28. The Lazarus Experiment (2007) 29. The Doctor's Daughter (2008) 30. Amy's Choice (2010) 31. Vincent and the Doctor (2010) 32. The Doctor's Wife (2011) 33. Let's Kill Hitler (2011) 34. The Wedding of River Song (2011) 35. The Doctor, the Widow and the Wardrobe (2011) 36. The Name of the Doctor (2013) 37. The Day of the Doctor (2013) 38. The Time of the Doctor (2013) 39. The Husbands of River Song (2015)

 * We've not included The Next Doctor (2008) as the title refers to someone who turns out not to be the Doctor.
Also, we've not included The Talons of Weng-Chiang (1977) as 'Weng-Chiang' is an alias of Magnus Greel.

# The Timeline of River Song

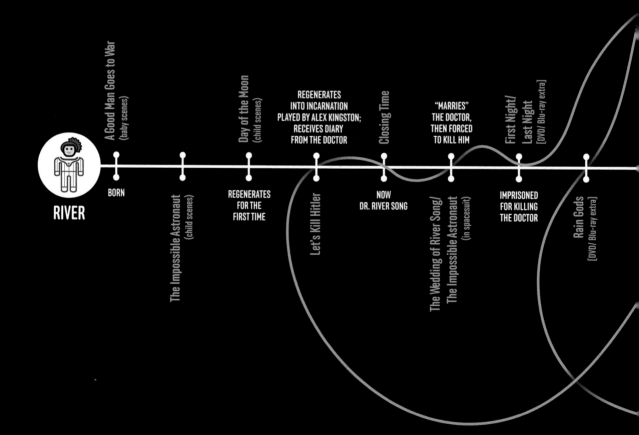

**RIVER**

**A Good Man Goes to War**
(baby scenes)

**BORN**

**The Impossible Astronaut**
(child scenes)

**Day of the Moon**
(child scenes)

**REGENERATES FOR THE FIRST TIME**

**REGENERATES INTO INCARNATION PLAYED BY ALEX KINGSTON; RECEIVES DIARY FROM THE DOCTOR**

**Let's Kill Hitler**

**Closing Time**

**NOW DR. RIVER SONG**

**"MARRIES" THE DOCTOR, THEN FORCED TO KILL HIM**

**The Wedding of River Song/ The Impossible Astronaut**
(in spacesuit)

**First Night/ Last Night**
[DVD/ Blu-ray extra]

**IMPRISONED FOR KILLING THE DOCTOR**

**Rain Gods**
[DVD/ Blu-ray extra]

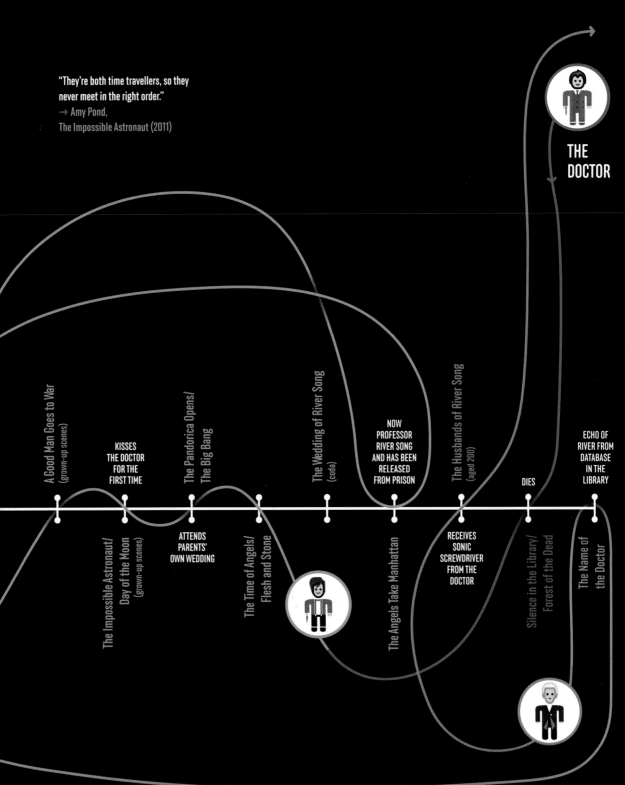

"They're both time travellers, so they never meet in the right order."
→ Amy Pond,
The Impossible Astronaut (2011)

THE
DOCTOR

A Good Man Goes to War
(grown-up scenes)

KISSES
THE DOCTOR
FOR THE
FIRST TIME

The Impossible Astronaut/
Day of the Moon
(grown-up scenes)

ATTENDS
PARENTS'
OWN WEDDING

The Pandorica Opens/
The Big Bang

The Time of Angels/
Flesh and Stone

The Wedding of River Song
(coda)

The Angels Take Manhattan

NOW
PROFESSOR
RIVER SONG
AND HAS BEEN
RELEASED
FROM PRISON

RECEIVES
SONIC
SCREWDRIVER
FROM THE
DOCTOR

The Husbands of River Song
(aged 200)

DIES

Silence in the Library/
Forest of the Dead

ECHO OF
RIVER FROM
DATABASE
IN THE
LIBRARY

The Name of
the Doctor

# THE SIXTH DOCTOR

## VITAL STATISTICS

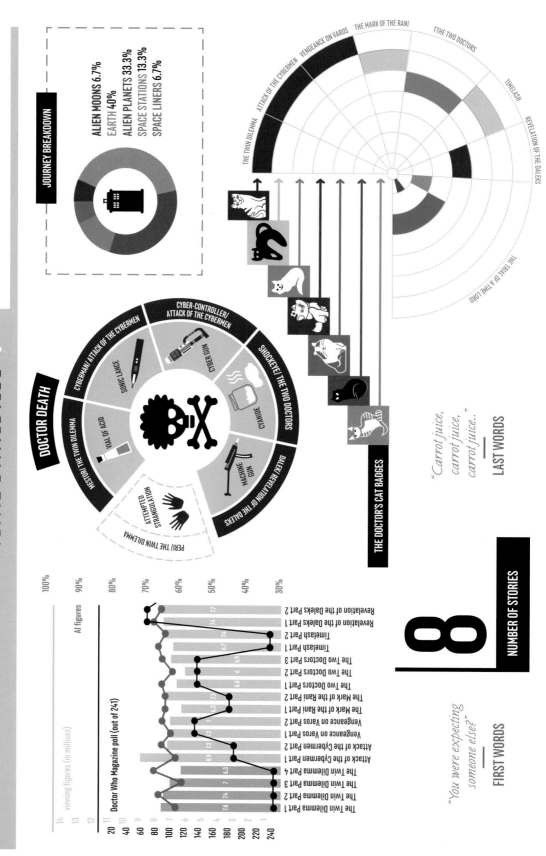

### JOURNEY BREAKDOWN

- **ALIEN MOONS** 6.7%
- **EARTH** 40%
- **ALIEN PLANETS** 33.3%
- **SPACE STATIONS** 13.3%
- **SPACE LINERS** 6.7%

### DOCTOR DEATH

- MESTOR/ THE TWIN DILEMMA — VIAL OF ACID
- CYBERMAN/ ATTACK OF THE CYBERMEN — SONIC LANCE
- CYBER-CONTROLLER/ ATTACK OF THE CYBERMEN — CYBER GUN
- SHOCKEYE/ THE TWO DOCTORS — CYANIDE
- DALEK/ REVELATION OF THE DALEKS — MACHINE GUN
- PERI/ THE TWIN DILEMMA — ATTEMPTED STRANGULATION

### THE DOCTOR'S CAT BADGES

Stories (outer ring): THE TWIN DILEMMA, ATTACK OF THE CYBERMEN, VENGEANCE ON VAROS, THE MARK OF THE RANI, THE TWO DOCTORS, TIMELASH, REVELATION OF THE DALEKS, THE TRIAL OF A TIME LORD

### LAST WORDS

*"Carrot juice, carrot juice, carrot juice..."*

### FIRST WORDS

*"You were expecting someone else?"*

### NUMBER OF STORIES

# 8

### Doctor Who Magazine poll (out of 241)

Viewing figures (in millions) — All figures

| Story | |
|---|---|
| The Twin Dilemma Part 1 | 7.6 |
| The Twin Dilemma Part 2 | 7.4 |
| The Twin Dilemma Part 3 | 7 |
| The Twin Dilemma Part 4 | 6.3 |
| Attack of the Cybermen Part 1 | 8.9 |
| Attack of the Cybermen Part 2 | 7.2 |
| Vengeance on Varos Part 1 | 7 |
| Vengeance on Varos Part 2 | 7.2 |
| The Mark of the Rani Part 1 | 6.3 |
| The Mark of the Rani Part 2 | 7.3 |
| The Two Doctors Part 1 | 6.6 |
| The Two Doctors Part 2 | 6 |
| The Two Doctors Part 3 | 6.9 |
| Timelash Part 1 | 6.7 |
| Timelash Part 2 | 7.4 |
| Revelation of the Daleks Part 1 | 7.4 |
| Revelation of the Daleks Part 2 | 7.7 |

## APPEARANCES IN OTHER STORIES

TIME AND THE RANI (1987)
**THE NEXT DOCTOR (2008)**
THE ELEVENTH HOUR (2010)
**THE NAME OF THE DOCTOR (2013)**
THE DAY OF THE DOCTOR (2013)

19%
81%

PERI BROWN
25 EPISODES

MEL BUSH
6 EPISODES

## HONOURABLE MENTIONS

SLIPBACK (RADIO PLAY, 1985)
THE ULTIMATE ADVENTURE
(STAGE PLAY, 1989)

## FASHION SHOW

# 31
**NUMBER OF EPISODES**

100%
90%
80%
70%
60%
50%
40%
30%

viewing figures (in millions)

14
13
12
11
10
9
8
7
6
5
4
3
2
1

20
40
60
80
100
120
140
160
180
200
220
240

Doctor Who Magazine poll (out of 241)

All figures

The Trial of a Time Lord Part 1 — 4.9
The Trial of a Time Lord Part 2 — 4.9
The Trial of a Time Lord Part 3 — 3.9
The Trial of a Time Lord Part 4 — 3.7
The Trial of a Time Lord Part 5 — 4.8
The Trial of a Time Lord Part 6 — 4.6
The Trial of a Time Lord Part 7 — 5.1
The Trial of a Time Lord Part 8 — 5
The Trial of a Time Lord Part 9 — 5.2
The Trial of a Time Lord Part 10 — 4.6
The Trial of a Time Lord Part 11 — 5.3
The Trial of a Time Lord Part 12 — 5.3
The Trial of a Time Lord Part 13 — 4.4
The Trial of a Time Lord Part 14 — 5.6

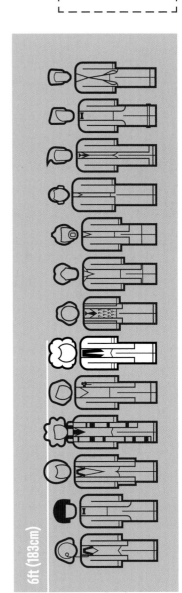

6ft (183cm)

6
5
4
3
2
1

# How to Kill a Time Lord

**The Doctor and his people can live for ever, barring accidents or getting...**

...all the energy of the Time Vortex inside him or her[1]; atomised in a dispersal chamber[2]; blasted by astronaut's gauntlet[3]; blasted by Dalek gun[4]; blasted by Rassilon's gauntlet[5]; blasted with a ray gun[6]; blasted with psychic spider electricity[7]; bored with current incarnation[8]; burnt up as fuel[9]; caught inside a crashing space or time craft[10]; cut off from the Matrix while connected to it[11]; decontaminated of ape bacteria and germs[12]; dropped from a great height[13]; drowned[14]; eaten by Reaper[15]; erased from time and space[16]; fed aspirin[17]; forced to regenerate by other Time Lords[18]; miniaturised[19]; old and tired[20]; operated on by human surgeon[21]; poisoned by radiation[22]; poisoned by Spectrox toxaemia[23]; poisoned with lipstick[24]; shot, then deciding not to regenerate[25]; stabbed in the back[26]; through all 12 regenerations[27]; transformed into anti-matter while in another universe, then coming back into our universe[28]; used for spare parts[29]; zapped by thunderbolts[30].

---

1       The Ninth Doctor, The Parting of the Ways (2005)
2       Morbius, The Brain of Morbius (1976)
3       The Eleventh Doctor, The Impossible Astronaut (2011)
4       The Master, Doctor Who (1996); the Tenth Doctor, The Stolen Earth (2008)
5       Partisan, The End of Time (2009–10)
6       The War Chief, The War Games (1969); numerous others in subsequent stories.
7       K'Anpo, Planet of the Spiders (1974)
8       Romana, Destiny of the Daleks (1979)
9       The Twelfth Doctor, Heaven Sent (2015)
10      The Sixth Doctor, Time and the Rani (1987); the Eighth Doctor, The Night of the Doctor (2013)
11      Goth, The Deadly Assassin (1976)
12      Almost kills the Eleventh Doctor, Cold Blood (2010)
13      Morbius, The Brain of Morbius; the Fourth Doctor, Logopolis (1981)
14      The Tenth Doctor, Turn Left (2008)
15      The Ninth Doctor, Father's Day (2005)
16      The Second Doctor is threatened with this, The War Games (1969)
17      The Third Doctor says one could kill him, The Mind of Evil (1971)
18      The Second Doctor, The War Games (1969)
19      Camera technician, The Deadly Assassin (1976)
20      The First Doctor, The Tenth Planet (1966); the War Doctor, The Day of the Doctor (2013);
        the Eleventh Doctor, The Time of the Doctor (2013)
21      The Seventh Doctor, Doctor Who (1996)
22      The Third Doctor, Planet of the Spiders; the Tenth Doctor, The End of Time (2009–10)
23      The Fifth Doctor, The Caves of Androzani (1984)
24      Almost kills the Eleventh Doctor, Let's Kill Hitler (2011)
25      The Master, Last of the Time Lords (2007)
26      Runcible, The Deadly Assassin (1976)
27      Azmael, The Twin Dilemma (1984)
28      Omega, Arc of Infinity (1983)
29      The Corsair and other Time Lords, The Doctor's Wife (2011)
30      Two Time Lords, The Five Doctors (1983)

"I can still die. If I'm killed before regeneration, then I'm dead. Even then, even if I change, it feels like dying. Everything I am dies. Some new man goes sauntering away, and I'm dead."

→ The Tenth Doctor, The End of Time (2009–10)

THE
RESCUE
(1965)

PYRAMIDS
OF
MARS
(1975)

HORROR
OF
FANG
ROCK
(1977)

ATTACK
OF THE
CYBERMEN
(1985)

THE
PARTING
OF THE
WAYS
(2005)

THE
DOCTOR'S
WIFE
(2011)

*In Deepest Sympathy*

*The Edge of Destruction* (1964) features only the Doctor and his companions anyway. Apart from the Doctor and his companions, in *The Caves of Androzani* (1985) only Timmin survives and in *The God Complex* (2011) only Gibbis survives.

"Pain and loss, they define us as much as happiness or love. Whether it's a world, or a relationship, everything has its time. And everything ends."
→ Sarah Jane Smith, School Reunion (2006)

# EVERYBODY Dies!!

In 6 stories, all seen, speaking characters die except the Doctor and his companions.

# WEIRD WEAPONS

Total destruction – in the strangest ways!

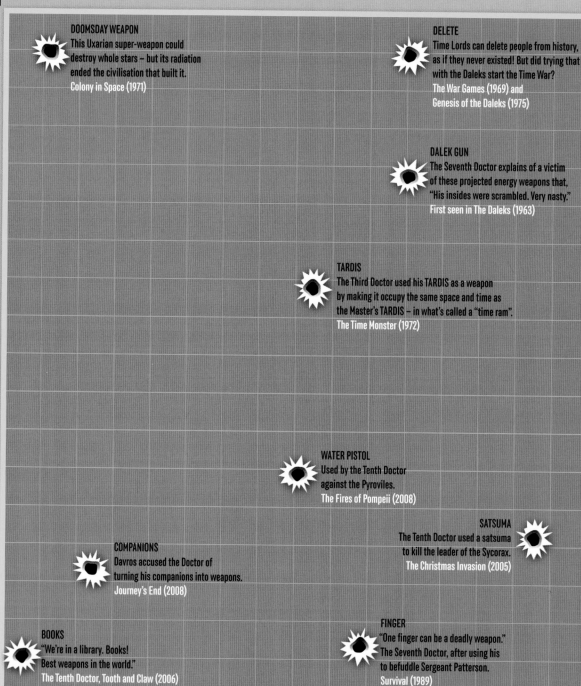

**HIGH**

**DESTRUCTIVENESS**

**DOOMSDAY WEAPON**
This Uxarian super-weapon could destroy whole stars – but its radiation ended the civilisation that built it.
Colony in Space (1971)

**DELETE**
Time Lords can delete people from history, as if they never existed! But did trying that with the Daleks start the Time War?
The War Games (1969) and Genesis of the Daleks (1975)

**DALEK GUN**
The Seventh Doctor explains of a victim of these projected energy weapons that, "His insides were scrambled. Very nasty."
First seen in The Daleks (1963)

**TARDIS**
The Third Doctor used his TARDIS as a weapon by making it occupy the same space and time as the Master's TARDIS – in what's called a "time ram".
The Time Monster (1972)

**WATER PISTOL**
Used by the Tenth Doctor against the Pyroviles.
The Fires of Pompeii (2008)

**SATSUMA**
The Tenth Doctor used a satsuma to kill the leader of the Sycorax.
The Christmas Invasion (2005)

**COMPANIONS**
Davros accused the Doctor of turning his companions into weapons.
Journey's End (2008)

**BOOKS**
"We're in a library. Books! Best weapons in the world."
The Tenth Doctor, Tooth and Claw (2006)

**FINGER**
"One finger can be a deadly weapon."
The Seventh Doctor, after using his to befuddle Sergeant Patterson.
Survival (1989)

**LOW**

**WEIRDNESS**

**THE DEAD**
Missy helped turn dead people
into an army of Cybermen.
Dark Water / Death in Heaven (2014)

**REALITY BOMB**
Davros used 27 stolen planets
to power this weapon that would
destroy all matter in every universe.
Journey's End (2008)

**STATUE**
The Doctor used a statue made
of living metal, validium, to destroy
a fleet of Cyber warships.
Silver Nemesis (1988)

**ANTIMATTER**
The Cybermen turned a spaceship
powered by antimatter into a weapon
by setting it on a collision course with Earth.
Earthshock (1982)

**THE MOMENT**
A strange Time Lord weapon, the
Moment can see the future and tell
you whether using its power will work!
The Day of the Doctor (2013)

**THE DYING**
Davros converted people on the
point of death in a funeral parlour
into an army of Daleks.
Revelation of the Daleks (1985)

**PLASTIC**
The Nestene Consciousness can
animate any plastic as a weapon –
including shop window dummies,
a chair, a doll, plastic flowers and
the flex on the Doctor's telephone.
First seen in Spearhead from Space (1970)

**HUMAN FACTOR**
The Second Doctor starts a civil war
among the Daleks by infecting some
of them with human qualities
such as courage, mercy and friendship.
The Evil of the Daleks (1967)

**BOW SHIPS**
Swift Time Lord craft that killed
giant vampires by firing bolts
of steel through their hearts.
State of Decay (1980)

**SONIC GUN**
The weapons of the Ice Warriors,
"will burst your brain with noise."
First seen in The Ice Warriors (1967)

**VINEGAR**
The Ninth Doctor and his friends
defeated a Slitheen invasion of
Earth using vinegar – and missiles.
World War Three (2005)

**SINGING**
Chimerons make a noise that
is partly a song and partly
a defence mechanism.
Delta and the Bannermen (1987)

**TISSUE COMPRESSION ELIMINATOR**
The Master used this to kill
people – and shrink their bodies
down to the size of dolls.
First used in Terror of the Autons (1971)

**SONIC BLASTER**
A 51st century weapon used by
both Captain Jack and River Song
to make square holes in walls.
First seen in The Doctor Dances (2005)

**WOODEN HORSE**
The First Doctor suggests the wooden
horse to the Greeks to help them win
the siege of Troy – though he thinks
the idea is "obviously absurd".
The Myth Makers (1965)

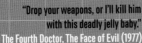
**JELLY BABY**
"Drop your weapons, or I'll kill him
with this deadly jelly baby."
The Fourth Doctor, The Face of Evil (1977)

**HIGH**

# BOOC

Destruction wrought by Ace in the time she knows the Seventh Doctor, and vice versa.

## DRAGONFIRE (1987)
Pours milkshake on two people.
Destroys a frozen door with Nitro 9.

**1/10**
DESTRUCTIVENESS SCORE

## REMEMBRANCE OF THE DALEKS (1988)
Headbutts headmaster in the stomach.
Destroys Dalek with anti-tank rocket.
Damages two Daleks with augmented baseball bat.
Smashes through a window.
Tries to attack Mike on discovering his betrayal.

**7/10**
DESTRUCTIVENESS SCORE

## THE HAPPINESS PATROL (1988)
Tries to attack Priscilla P.
Knocks Priscilla P over.
Wounds Fifi with Nitro 9.
Threatens Kandyman with poker.

**2/10**
DESTRUCTIVENESS SCORE

## SILVER NEMESIS (1988)
Breaks into Windsor Castle.
Blows up Cyberman spaceship with Nitro 9.
Kills three Cybermen and wounds the Cyber Leader using gold coins.
Tricks two Cybermen into shooting each other.

**8/10**
DESTRUCTIVENESS SCORE

## THE GREATEST SHOW IN THE GALAXY (1988)
Smacks Bellboy's robot with shovel.
Tears hole in circus tent.
Threatens robots with a robot arm.
Hits and kicks robot bus conductor, knocks its hat off.
Uses Bellboy's robot to kill four clown robots and the Chief Clown.

**8/10**
DESTRUCTIVENESS SCORE

TOTAL
47/90

## BATTLEFIELD (1989)
Recounts blowing up an art room at school.
Blows up archaeological dig with Nitro 9 – sooner than the timer was set for.
Pulls Excalibur from stone – causing defence systems to attack.
Helps Shou Yuing (under influence of Morgaine's magic).
Uses hate speech while fighting Morgaine.
Stumbles into Morgaine.
With the Brigadier, blows up Arthur's spaceship.

**7/10**
DESTRUCTIVENESS SCORE

## GHOST LIGHT (1989)
Tries to jump on Redvers Fenn Cooper (but misses).
Threatens two husks with a cane.
Fighting Nimrod, breaks the membrane of a hibernation chamber.
Threatens Josiah with a gun (really a radiation detector).
Tries to scratch Josiah's face.
Fights with Gwendoline – twice.
Admits she burnt down Gabriel Chase in 1983;
wishes she'd blown it up instead.

**5/10**
DESTRUCTIVENESS SCORE

## THE CURSE OF FENRIC (1989)
Causes trouble when she inadvertently tells Dr Judson what new Viking engraving is for.
Causes trouble when she inadvertently tells Dr Judson what new Viking engraving is for.
Battles Haemovores.
With Doctor, steals chess set (her destructive impulses save herself and the Doctor from a trap).
Blows up wall with Nitro 9.
Causes trouble when she inadvertently picks up oriental flask containing Fenric.
Repels Ancient Haemovore with her faith in the Doctor to chess puzzle.

**5/10**
DESTRUCTIVENESS SCORE

## SURVIVAL (1989)
Fights back against Cheetah People – but trip wire doesn't work and snare only catches the Doctor.
Tries to fight Cheetah People.
Throws a rock and hits Karra's head.
Destructive impulses lead to her partially becoming a Cheetah Person.
Tempted to hunt.
Ready to duel Midge on a motorbike.

**4/10**
DESTRUCTIVENESS SCORE

124

**OOM!**

"Can I have a go, Professor?"
"Wanton destruction of public property? Certainly not."
→ Ace and the Seventh Doctor, The Happiness Patrol (1988)

### DRAGONFIRE (1987)
Talks dragon into opening its head – which leads to its death.
Tells Kane his people are dead – which leads to Kane's suicide. .

**1/10**
DESTRUCTIVENESS SCORE

### REMEMBRANCE OF THE DALEKS (1988)
Destroys Dalek with Nitro 9.
Destroys a Dalek mid-transmat and overloads the transmat machine.
Hurts own hand with augmented baseball bat.
Confuses three Daleks so they can be destroyed.
Smashes transmat with baseball bat.
Puts time controller out of phase.
Breaks into Dalek shuttle.
Shorts out a Dalek.
Uses Hand of Omega to vaporise Skaro's sun and destroy Skaro,
at least four other planets and the Dalek mothership with 400 Daleks on board.
Talks a Dalek into self-destructing.

**10/10**
DESTRUCTIVENESS SCORE

### THE HAPPINESS PATROL (1988)
Tricks Kandyman into losing his temper and breaking a bottle of lemonade.
Again uses lemonade to glue Kandyman to the floor.
Act of public disobedience causes argument between groups of the Happiness Patrol.
Uses Earl Sigma's harmonica to cause an avalanche that kills Fifi.
Threatens Kandyman with poker and ignites oven to chase him away with flames.
Makes Helen A cry.

**6/10**
DESTRUCTIVENESS SCORE

### GHOST LIGHT (1989)
Threatens Josiah with a gun (really a radiation detector).
Directs jet of steam into Josiah's face.
Psychically banishes Light from room.
Bamboozles Light into dissipating.

**4/10**
DESTRUCTIVENESS SCORE

**TOTAL 54/90**

### THE GREATEST SHOW IN THE GALAXY (1988)
Confuses robot bus conductor so it breaks, then shoots it with its ticket machine.
Hits and knocks out a robot clown with a juggling club.
Uses amulet to reflect energy back at three Gods of Ragnarok which kills them.
Throws amulet which explodes.
Destroys circus tent.

**8/10**
DESTRUCTIVENESS SCORE

### SILVER NEMESIS (1988)
Breaks into Windsor Castle.
Jams Cyberman signals with jazz.
Wakes Nemesis, causing an explosion.
Breaks Cybermen with the exhausts of a rocket sled.
Kills two Cybermen with the exhausts of a rocket sled.
Destroys the Cyber fleet of 1,000 warships using Nemesis.

**10/10**
DESTRUCTIVENESS SCORE

### BATTLEFIELD (1989)
Bursts an empty bang of crisps.
Scrapes roof of tunnel, making dust fall.
Breaks spaceship defence systems to free Ace.
Bamboozles Pat Rawlinson and Peter Warmsly into UNIT truck.
Drives Bessie so fast it burns rubber.
Threatens to decapitate Mordred (a bluff).
Taunts Morgaine into releasing the Destroyer.
Prepares to shoot the Destroyer.
Pulls helmet off dead King Arthur.
Takes hold of Mordred's arm then knocks him unconscious using fingers.

**5/10**
DESTRUCTIVENESS SCORE

### THE CURSE OF FENRIC (1989)
Battles Haemovores, reciting names of his companions to repel them.
With Ace, steals chess set.
Breaks Ace's faith in him.
Convinces Ancient Haemovore to kill itself and Fenric.

**5/10**
DESTRUCTIVENESS SCORE

### SURVIVAL (1989)
Bamboozles Sergeant Patterson using one finger.
Trips up a Cheetah Person with umbrella.
Knocks Ace to the ground and duels Midge on a motorbike,
leading to explosion and Midge being badly wounded.
Fights the Master.
Destructive impulses lead to him partially becoming a Cheetah Person.
Fight with the Master destroys the planet of the Cheetah People.

**5/10**
DESTRUCTIVENESS SCORE

# THE OTHER DOCTOR:
## VITAL STATISTICS

## THE WATCHER

### 4
NUMBER OF EPISODES

### 0
NUMBER OF LINES

"He was the Doctor all along."
→ Nyssa, Logopolis (1981)

### THE WATCHER'S POSES

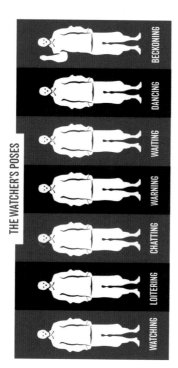

WATCHING | LOITERING | CHATTING | WARNING | WAITING | DANCING | BECKONING

## THE VALEYARD

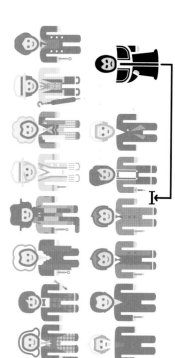

"The Valeyard is an amalgamation of the darker sides of your nature, somewhere between your twelfth and final incarnation..."
→ The Master, The Trial of a Time Lord (1986)

**valeyard** /ˈvælˈjɑːd/
*n.* learned court prosecutor
[from THE CONCISE
GALLIFREYAN DICTIONARY]

### 14
NUMBER OF EPISODES

## JACKSON LAKE

**TARDIS**
Tethered Aerial
Release Developed
In Style

### JACKSON'S 'SONIC' SCREWDRIVER

**FAMILY TREE**

JACKSON ═ CAROLINE

FREDERICK

"You became the Doctor because
the infostamp you picked up was
a book about one particular man."
→ The Tenth Doctor, The Next Doctor (2008)

**1**

**NUMBER OF EPISODES**

---

---

## THE 'METACRISIS' TENTH DOCTOR

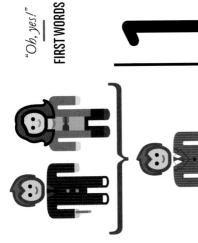

*"Oh, yes!"*
**FIRST WORDS**

"I'm unique.
Never been
another like me."
→ The Metacrisis Doctor,
Journey's End (2008)

**1**
**NUMBER OF HEARTS**

**1**
**NUMBER OF EPISODES**

---

## THE WAR DOCTOR

*"Doctor no more."*
**FIRST WORDS**

**OTHER APPEARANCES**
LISTEN (2014)
THE ZYGON INVASION (2015)

*"I hope the ears
are a bit less
conspicuous this time."*
**LAST WORDS**

**2**
**+1**
**MINI-EPISODE**

**NUMBER OF EPISODES**

---

## THE DREAM LORD

**1**

**NUMBER OF EPISODES**

"The Dream Lord was me. Psychic pollen.
It's a mind parasite. It feeds on everything
dark in you, gives it a voice, turns it against you."
→ The Eleventh Doctor, Amy's Choice (2010)

# THE SEVENTH DOCTOR

## VITAL STATISTICS

### PET PEEVES

BURNT TOAST • BUS STATIONS
UNREQUITED LOVE • TYRANNY • CRUELTY

### SKILLS

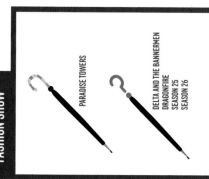

PLAYING THE SPOONS

JUGGLING

MAGIC TRICKS

ESCAPOLOGY

RIDING A MOTORCYCLE, DRIVING A VAN/ BUGGY/ CAR

### NUMBER OF LITERAL CLIFFHANGERS

11

### FASHION SHOW

PARADISE TOWERS

DELTA AND THE BANNERMEN
DRAGONFIRE
SEASON 25
SEASON 26

### LAST WORDS

*"Timing malfunction!*
*The Master, he's out there...*
*I know.. I've got to stop.. him..."*

### FIRST WORDS

*"Oh no, Mel!"*

### The Doctor's Malapropisms

Awork wides the nose grow louder
Every Dogfish has its day
Two wrongs don't make a left turn
Time and tide wait for no man
A stitch in time takes up space

### SEVENTH DOCTOR
`<<< TARDIS LOG >>>`

LAKERTYA ■ PARADISE TOWERS
TOLL PORT 5715 ■
SHANGRI-LA HOLIDAY CAMP,
SOUTH WALES, 1950s
ICEWORLD ON SVARTOS
LONDON, EARTH 1963 ■ TERRA ALPHA
WINDSOR, EARTH, 1988 and 1638
SEGONAX ■ LAKE VORTIGERN, EARTH
NORTHUMBRIA, EARTH, 1940s
PERIVALE, EARTH
SAN FRANCISCO, EARTH, 1999

### NUMBER OF EPISODES

42

#### viewing figures (in millions)
#### Doctor Who Magazine poll (out of 241)

AI figures

| | viewing figures | AI figures |
|---|---|---|
| Time and the Rani Part 1 | 5.1 | |
| Time and the Rani Part 2 | 4.2 | |
| Time and the Rani Part 3 | 4.3 | |
| Time and the Rani Part 4 | 4.9 | |
| Paradise Towers Part 1 | 4.5 | |
| Paradise Towers Part 2 | 5.2 | |
| Paradise Towers Part 3 | 5 | |
| Paradise Towers Part 4 | 5 | |
| Delta and the Bannermen Part 2 | 5.3 | |
| Delta and the Bannermen Part 3 | 5.1 | |
| Dragonfire Part 1 | 5.5 | 5.4 |
| Dragonfire Part 2 | 5 | |
| Dragonfire Part 3 | 4.7 | |

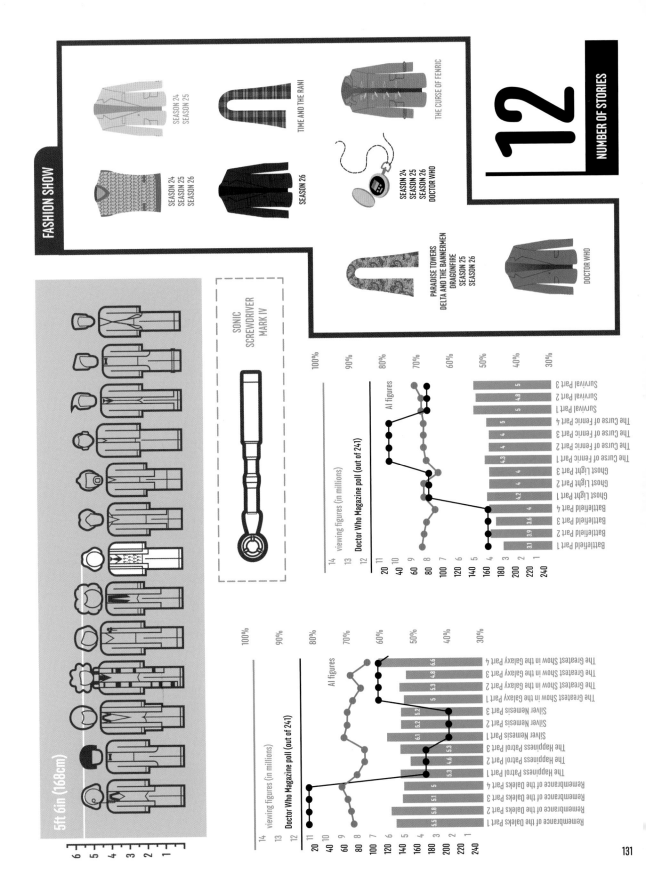

SEASON 24
SEASON 25

SEASON 24
SEASON 25
SEASON 26

SEASON 26

TIME AND THE RANI

THE CURSE OF FENRIC

SEASON 24
SEASON 25
SEASON 26
DOCTOR WHO

PARADISE TOWERS
DELTA AND THE BANNERMEN
DRAGONFIRE
SEASON 25
SEASON 26

DOCTOR WHO

12
**NUMBER OF STORIES**

5ft 6in (168cm)

SONIC SCREWDRIVER MARK IV

14 viewing figures (in millions)
13
12 Doctor Who Magazine poll (out of 241)

All figures

100%
90%
80%
70%
60%
50%
40%
30%

Battlefield Part 1 — 3.1
Battlefield Part 2 — 3.9 3.6
Battlefield Part 3 — 4
Battlefield Part 4 — 4.2
Ghost Light Part 1 — 4
Ghost Light Part 2 — 4
Ghost Light Part 3 — 4.3
The Curse of Fenric Part 1 — 4
The Curse of Fenric Part 2 — 4
The Curse of Fenric Part 3 — 5
The Curse of Fenric Part 4 — 5
Survival Part 1 — 4.8
Survival Part 2 — 5
Survival Part 3

20 40 60 80 100 120 140 160 180 200 220 240

14 viewing figures (in millions)
13
12 Doctor Who Magazine poll (out of 241)

All figures

100%
90%
80%
70%
60%
50%
40%
30%

Remembrance of the Daleks Part 1
Remembrance of the Daleks Part 2 — 5.8 5.1
Remembrance of the Daleks Part 3 — 5.5
Remembrance of the Daleks Part 4 — 5.3
The Happiness Patrol Part 1 — 4.6 5.3
The Happiness Patrol Part 2
The Happiness Patrol Part 3 — 6.1 5.2
Silver Nemesis Part 1 — 5
Silver Nemesis Part 2 — 5.3 4.8 6.6
Silver Nemesis Part 3
The Greatest Show in the Galaxy Part 1
The Greatest Show in the Galaxy Part 2
The Greatest Show in the Galaxy Part 3
The Greatest Show in the Galaxy Part 4

20 40 60 80 100 120 140 160 180 200 220 240

6 5 4 3 2 1

# The Sonic Screwdriver

"Harmless is just the word: that's why I like it! Doesn't kill, doesn't wound, doesn't maim. But I'll tell you what it does do: it is very good at opening doors."

→ The Tenth Doctor, Doomsday (2006)

Repairing organic parts • Operating computers • Creating fire • Overloading robot sensors • Bouncing waves off a knife • Cutting through a wall • Scanning for alarm systems • Detonation of landmines • Unlocking a holding cell • Operating cash machines • Opening electronic doors • Creating sparks • Unbolting doors • Fusing shut a sliding door • Opening a lift door • Distracting giant maggots • Detecting booby-trapped floor tiles • Breaking a hypnotic trance • Opening a refinery door • Unlocking a taxi door and window • Remotely detonating mines • Cutting locks • Tightening and loosening screws • Repairing wires • Fixing a circle of transmat refractors • Breaching a force field • Sabotaging a two-way radio • Deactivating an energy loop • Safe-cracking • Unlocking handcuffs • Hacking into websites • Turning off security systems • Locking and unlocking hatches • Disabling emergency exit alarms and locks • Fusing locks shut • Opening bus doors • Unlocking handcuffs • Acting as a medical scanner • Partly reversing the Abzorbaloff's absorption of Ursula Blake • Building a DNA scanning device • Scanning for fluctuating DNA • Detecting and stopping telepathic signals • Scanning a life form for information • Destroying the controls of a lift • Establishing a computer interface • Creating a hole in a force field • Blowing up a Dalek bomb • Creating loud noises •

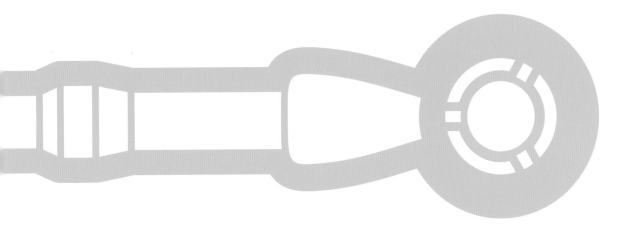

Reversing magnetic fields •
Performing TARDIS maintenance •
Scanning for life signs • Calculations •
Calling phone numbers • Detonating an explosive device • Searching a smartphone for an app • Dematerialising the TARDIS from afar • Activating ventilation air ducts
• Changing the destination of a teleport • Tinting glass • Activating the memory-erasing device in the Black Archive • Disabling a shimmer • Corroding barbed wire •
Cutting rope • Acting as a flashlight • Cutting a spider web • **THINGS WE'VE SEEN IT DO** • Uncorking a champagne bottle •
Analysing and healing wounds • Detecting differences between gangers and humans • Electrifying a dream crab's nerve centres • Scanning for heat signatures •
Scanning an apple • Shattering glass • Exploding lightbulbs • Hacking into computer records • Disabling force fields • Switching off CCTV monitors • Repairing lifts •
Deleting answerphone messages •
Changing train signals • Detecting the
location of a mobile phone

Unlock wooden doors •
Wound or kill living things •
Unlock a deadlock seal •
**AND THINGS
IT CAN'T DO**
• Stop giant wooden toys
• Survive a Terileptil's
laser gun

# MIXED MEDIA

Over 52 years, there have been huge changes in how we've received the Doctor's adventures in different media...

⚡ 3 October 1983 **Release of first VHS of a TV Doctor Who story**

⚡ 3 January 1970 **Broadcast of first colour episode of Doctor Who story (Revenge of the Cybermen (1975))**

⚡ 3 January 1970 Broadcast of first colour episode of Doctor Who (Spearhead from Space episode 1)

⚡ 12 November 1964 **Publication of first novelisation of a TV Doctor Who story**

⚡ 14 November 1964 **Publication of first comic strip Doctor Who (The Klepton Parasites part 1)**

⚡ 23 November 1963 **Broadcast of first episode of Doctor Who – in black and white (An Unearthly Child)**

⚡ 14 November 1964 Publication of first Doctor Who in an Exciting Adventure with the Daleks, adapted from The Daleks (1963–4))

⚡ 21 December 1965 **Opening night of first stage play based on Doctor Who (The Curse of the Daleks)**

⚡ April 1966 **Release of first audio soundtrack of a Doctor Who story (excerpts from The Chase)**

⚡ 11 October 1979 **Publication of first magazine devoted entirely to the programme – issue 1 of Doctor Who Weekly (now Doctor Who Magazine)**

1 November 1999 Release of first DVD of a TV Doctor Who story (The Five Doctors (1983))

24 November 2003 Release of final VHS of Doctor Who (existing episodes from The Reign of Terror (1964), The Faceless Ones (1967) and The Web of Fear (1968))

26 March 2005 First episode broadcast in 16:9 aspect ratio (Rose)

11 April 2009 Broadcast of first episode of Doctor Who in high definition (Planet of the Dead)

23 November 2013 Broadcast of first and to date final episode of Doctor Who in 3-D. Also first episode to be shown in cinemas at the same time as broadcast (The Day of the Doctor)

5 October 1988 Broadcast of first episode of Doctor Who in stereo sound (Remembrance of the Daleks part 1)

13 November 2003 Webcast of first fully animated Doctor Who story, Scream of the Shalka episode 1

25 December 2007 First episode of Doctor Who made available on BBC iPlayer for streaming or downloading (Voyage of the Damned)

29 June 2009 Release of first Blu-ray of a Doctor Who episode (Planet of the Dead)

Autumn 1983 Release of first Doctor Who computer game (The First Adventure)

135

# Type Forte

The first use of particular letters and punctuation in Doctor Who episode titles

25 JULY 1964
FIRST ONE-WORD EPISODE TITLE: KIDNAP

19 MAY 2007
FIRST EPISODE TITLE WITHOUT ANY LETTERS: 42

CHAPTER FOUR
“THAT’S SILLY”

CHAPTER NINE
DALEKS

CHAPTER SEVEN
TECHNOLOGY

CHAPTER ELEVEN
ALIEN WORLDS

APPENDIX
THE MAKING OF WHOGRAPHICA

CHAPTER EIGHT
MIND THE GAP

CHAPTER TWELVE
16 DAYS,
8 HOURS,
13 MINUTES,
51 SECONDS...
AND COUNTING

# THE EIGHTH DOCTOR
## VITAL STATISTICS

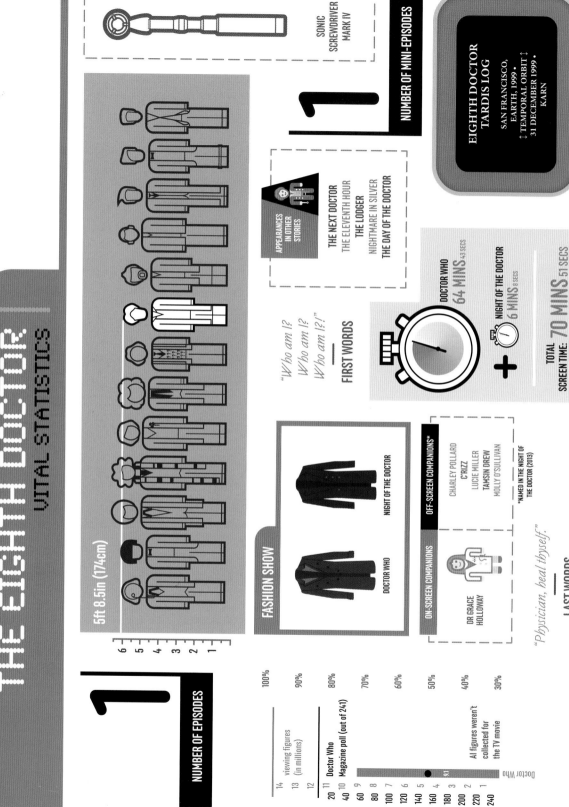

5ft 8.5in (174cm)

**NUMBER OF EPISODES**

1

**NUMBER OF MINI-EPISODES**

1

**SONIC SCREWDRIVER MARK IV**

**EIGHTH DOCTOR TARDIS LOG**
SAN FRANCISCO, EARTH, 1999 •
↑ TEMPORAL ORBIT ↑
31 DECEMBER 1999 •
KARN

**APPEARANCES IN OTHER STORIES**
THE NEXT DOCTOR
THE ELEVENTH HOUR
THE LODGER
NIGHTMARE IN SILVER
THE DAY OF THE DOCTOR

*"Who am I?*
*Who am I?*
*Who am I?!"*
**FIRST WORDS**

DOCTOR WHO
**64 MINS** 43 SECS

NIGHT OF THE DOCTOR
**6 MINS** 8 SECS

TOTAL SCREEN TIME: **70 MINS** 51 SECS

**FASHION SHOW**
DOCTOR WHO
NIGHT OF THE DOCTOR

**ON-SCREEN COMPANIONS**
DR GRACE HOLLOWAY

**OFF-SCREEN COMPANIONS***
CHARLEY POLLARD
C'RIZZ
LUCIE MILLER
TAMSIN DREW
MOLLY O'SULLIVAN

*NAMED IN THE NIGHT OF THE DOCTOR (2013)

*"Physician, heal thyself."*
**LAST WORDS**

viewing figures (in millions)
Doctor Who Magazine poll (out of 241)

100%
90%
80%
70%
60%
50%
40%
30%

14 | 20
13 | 40
12 | 60
11 | 80
10 | 100
9 | 120
8 | 140
7 | 160
6 | 180
5 | 200
4 | 220
3 | 240

9.1

All figures weren't collected for the TV movie

Doctor Who

# MIND THE GAP

The dark years without new Doctor Who on television.

## Survival Part 3

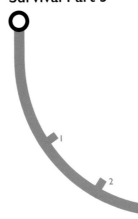

## Key

1 21 November 1990: BBC Two broadcasts Search Out Space,
an educational programme featuring the Seventh Doctor, Ace and K-9.

2 June 1991: Virgin Books publish the first in a range of original Doctor Who novels,
Timewyrm: Genesys, starring the Seventh Doctor. Original Doctor Who novels
have been published ever since by various companies.

3 27 August – 24 September 1993: BBC Radio 5 broadcasts
The Paradise of Death, starring the Third Doctor.

4 26–27 November 1993: BBC One broadcasts Dimensions in Time,
Doctor Who / EastEnders special for charity Children in Need,
starring the Third, Fourth, Fifth, Sixth and Seventh Doctors.

## Doctor Who

5     6   7   8

5 20 January – 24 February 1996: BBC Radio 2 broadcasts
The Ghosts of N-Space, starring the Third Doctor.

6 12 March 1999: BBC One broadcasts The Curse of Fatal Death,
a comedy serial for Comic Relief with various
celebrities playing the Doctor.

7 19 July 1999: Big Finish productions releases the first in a
range of original Doctor Who audio plays, The Sirens of Time,
starring the Fifth, Sixth and Seventh Doctors.

8 13 November 1999: BBC Two broadcasts The Web of Caves,
a comedy sketch with Mark Gatiss as the Doctor,
as part of Doctor Who Night.

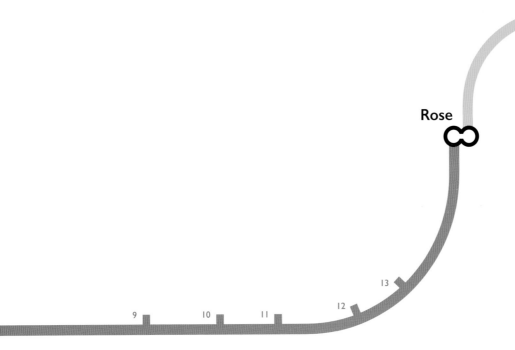

**Rose**

∞

9   12 July 2001: BBC Doctor Who website webcasts pilot episode of Death Comes to Time, starring the Seventh Doctor.

10   14 February – 3 May 2002: BBC Doctor Who website webcasts the remaining four episodes of Death Comes to Time.

11   2 August – 6 September 2002: BBC Doctor Who website webcasts Real Time, starring the Sixth Doctor.

12   2 May – 6 June 2003: BBC Doctor Who website webcasts Shada, starring the Eighth Doctor.

13   13 November – 18 December 2003: BBC Doctor Who website webcasts Scream of the Shalka, with Richard E Grant

Throughout this time, there were also new comic-strip adventures for the Doctor every four weeks in the pages of Doctor Who Magazine.

CHAPTER FOUR
"THAT'S SILLY"

**CHAPTER NINE**
**DALEKS**

CHAPTER SEVEN
TECHNOLOGY

CHAPTER ELEVEN
ALIEN WORLDS

APPENDIX
THE MAKING OF WHOGRAPHICA

CHAPTER EIGHT
MIND THE GAP

CHAPTER TWELVE
16 DAYS, 8 HOURS, 13 MINUTES, 51 SECONDS... AND COUNTING

# THE NINTH DOCTOR
## VITAL STATISTICS

"*Run!*"
——
FIRST WORDS

THE END OF THE WORLD 'AN INVITATION TO THE EARTH DEATH CELEBRATIONS – PLUS ONE' THE LONG GAME SATELLITE FIVE MANAGEMENT CREDENTIALS THE EMPTY CHILD DOCTOR JOHN SMITH, MINISTRY OF ASTEROIDS

→ PSYCHIC PAPER, THE ECCLESTON YEARS

# 13
**NUMBER OF EPISODES**

## KISSES

ROSE TYLER
CAPTAIN JACK HARKNESS

SONIC
SCREWDRIVER
MARK VI

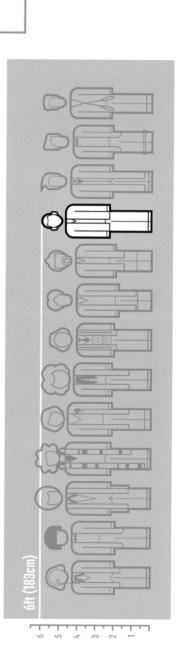

6ft (183cm)

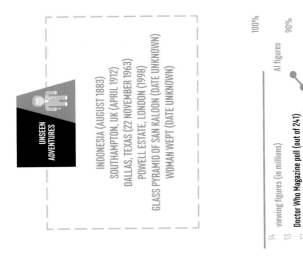

## V-NECK SHOW

ROSE
END OF THE WORLD
ALIENS OF LONDON
WORLD WAR THREE

THE UNQUIET DEAD

DALEK
THE LONG GAME
FATHER'S DAY

THE EMPTY CHILD/
THE DOCTOR DANCES
BOOM TOWN

BAD WOLF/
THE PARTING
OF THE WAYS

*"Rose.. before I go,*
*I just wanna tell you,*
*you were fantastic.*
*Absolutely fantastic.*
*And do you know what?*
*So was I!"*

### LAST WORDS

**UNSEEN ADVENTURES**

INDONESIA (AUGUST 1883)
SOUTHAMPTON, UK (APRIL 1912)
DALLAS, TEXAS (22 NOVEMBER 1963)
POWELL ESTATE, LONDON (1998)
GLASS PYRAMID OF SAN KALOON (DATE UNKNOWN)
WOMAN WEPT (DATE UNKNOWN)

AI figures

viewing figures (in millions)

**Doctor Who Magazine poll (out of 241)**

Rose — 10.8
The End of the World — 8
The Unquiet Dead — 8.9
Aliens of London — 7.6
World War Three — 8
Dalek — 8.6
The Long Game — 8.1
Father's Day — 7.1
The Empty Child — 6.9
The Doctor Dances — 7
Boom Town — 6.9
Bad Wolf — 6.8
The Parting of the Ways — 6.9

NINTH DOCTOR
TIME LOG

ROSE - MARCH 2005
THE END OF THE WORLD -
5.5/ APPLE/ 26
THE UNQUIET DEAD - 1869
ALIENS OF LONDON - MARCH 2006
WORLD WAR THREE - MARCH 2006
DALEK - 2012
THE LONG GAME - 200,000 AND 2012
FATHER'S DAY - 1987
THE EMPTY CHILD - 1941
THE DOCTOR DANCES - 1941
BOOM TOWN - 2006
BAD WOLF - 200,100
THE PARTING OF THE WAYS -
200,100 AND 2006

SHOUTING

TORTURED
POST-TIME WAR
ANGST

GOOFY
GRIN

**OTHER APPEARANCES**

ARMY OF GHOSTS
THE NEXT DOCTOR
THE ELEVENTH HOUR
THE LODGER
JOURNEY TO THE CENTRE OF THE TARDIS
NIGHTMARE IN SILVER
THE NAME OF THE DOCTOR

THE DAY OF THE DOCTOR

"RUN!"

"AND FOR
MY NEXT TRICK!"

# Episodes of the Daleks

They're his greatest foes, but how often does the Doctor actually battle them?

 **47** black and white episodes

**58** colour episodes

The Dead Planet •
The Survivors • The
Escape • The Ambush •
The Expedition • The Ordeal • The Rescue • World's End
• The Daleks • Day of Reckoning • The End of Tomorrow •
The Waking Ally • Flashpoint •
The Space Museum •The Final
Phase • The Executioners • The Death
of Time • Flight Through
Eternity • Journey into Terror •
The Death of Doctor Who • The Planet
of Decision • Mission to the Unknown •
The Nightmare Begins • Day of Armageddon • Devil's
Planet •The Traitors • Counter Plot • Coronas of the Sun
• Volcano • Golden Death • Escape Switch • The Abandoned Planet • The Destruction of Time • The Power of the Daleks Episode 1 •
The Power of the Daleks Episode 2 • The Power of the Daleks Episode 3 • The Power of the Daleks Episode 4 • The Power of the Daleks
Episode 5 • The Power of the Daleks Episode 6 • The Evil of the Daleks Episode 1 • The Evil of the Daleks
Episode 2 • The Evil of the Daleks Episode 3 • The Evil of the Daleks Episode 4 • The Evil of the Daleks
Episode 5 • The Evil of the Daleks Episode 6 • The Evil of the Daleks
Episode 7 • The War Games Episode 10 • The Mind of Evil Episode 3
(stock photo but new voice recording) • Day of the Daleks Episode 1 • Day
of the Daleks Episode 2 • Day of the Daleks Episode 3 • Day of the Daleks
Episode 4 • Frontier in Space Episode 6 • Planet of the Daleks Episode 1 • Planet of
the Daleks Episode 2 • Planet of the Daleks Episode 3 • Planet of the Daleks Episode 4
• Planet of the Daleks Episode 5 • Planet of the Daleks Episode 6 • Death to the
Daleks Part 1 • Death to the Daleks Part 2 • Death to the Daleks Part 3 • Death to the
Daleks Part 4 • Genesis of the Daleks Part 1 • Genesis of the Daleks Part 2 • Genesis of the
Daleks Part 3 • Genesis of the Daleks Part 4 • Genesis of the Daleks Part 5 • Genesis of the
Daleks Part 6 • Destiny of the Daleks Episode 1 • Destiny of the Daleks Episode 2 • Destiny of the
Daleks Episode 3 • Destiny of the Daleks Episode 4 • The Five Doctors • Resurrection of the Daleks
Part 1 • Resurrection of the Daleks Part 2 • Revelation of the Daleks Part 1 • Revelation of the Daleks Part
2 • Remembrance of the Daleks Part 1 • Remembrance of the Daleks Part 2 • Remembrance of the Daleks Part
3 • Remembrance of the Daleks Part 4 • Doctor Who (We don't see them but new voice recording) •
Dalek • Bad Wolf • The Parting of the Ways • Army of Ghosts • Doomsday • Daleks in Manhattan •
Evolution of the Daleks • The Stolen Earth • Journey's End • The Waters of Mars • The Beast Below (a shadow on wall) •
Victory of the Daleks • The Pandorica Opens • The Big Bang • The Wedding of River Song • Asylum of the Daleks • The Day of
the Doctor • The Time of the Doctor • Into the Dalek • The Magician's Apprentice • The Witch's Familiar • Hell Bent •

There are 105 Dalek episodes
(with new material recorded, not just clips) of 826.

12.71%

# Time of the Daleks
## Part 1

Frequency of words
used in Doctor Who
episode titles*

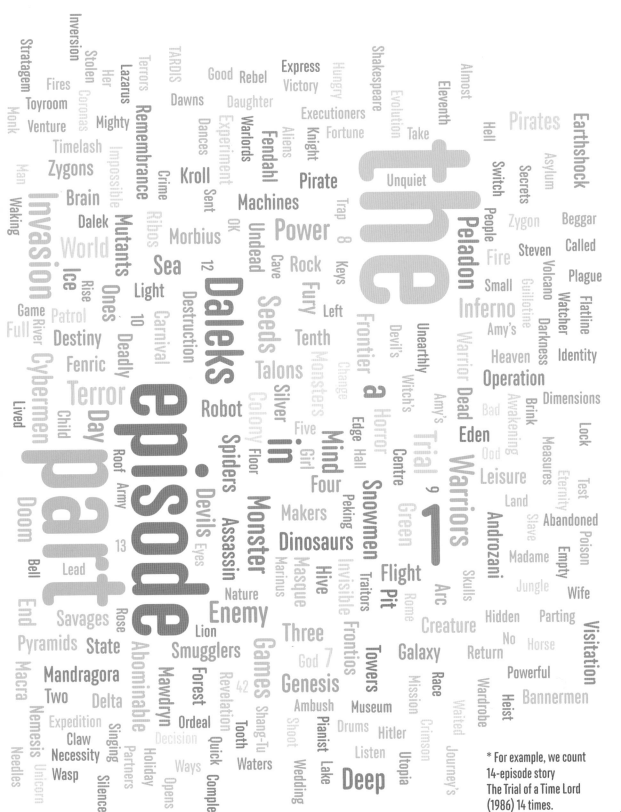

Inversion Stratagem Her Stolen Coronas Monk Fires Toyroom Venture Mighty Timelash Zygons Brain Man Waking Invasion World Ice Game Full Patrol River Destiny Fenric Lived Child Day Terror Cybermen Doom End Savages Bell Lead Pyramids State Macra Mandragora Two Delta Nemesis Expedition Needles Claw Necessity Singing Wasp Silence

Terrors TARDIS Lazarus Remembrance Dawns Good Rebel Express Victory Hungry Shakespeare Warlords Experiment Dances Fendahl Knight Fortune Eleventh Almost Hell Evolution Take Switch Asylum Pirates Earthshock Secrets Zygon Beggar Called

Dalek Rihos Crime Kroll Sent Machines Trap 8 Keys People Peladon Fire Steven Volcano Watcher Plague Small Guillotine Flatline Darkness Identity
Mutants OK Undead Cave Rock Left Inferno Amy's Heaven Operation Dimensions Lock Test
Light Rise Ones 12 Fury Tenth Devil's Unearthly Witch's Warrior Dead Brink Awakening Measures Eternity Abandoned Poison
Sea Destruction Five Floor Girl Change Amy's Ood Slave Empty
Robot Colony Silver Edge Hall Centre Green Warriors Androzani Madame Jungle Wife
Spiders Four Snowmen Land
Monster Makers Peking Traitors Flight Arc Skulls Hidden Parting No Horse Visitation
Enemy Masque Hive Pit Creature Return Powerful
Three Frontios Galaxy Race Bannermen Heist
Smugglers God 7 Towers Genesis Mission Crimson Utopia Journey's
Forest Revelation Ambush Museum Drums Hitler Waited Wardrobe
Mawdryn Ordeal Shang-Tu Pianist Lake Shoot Listen
Tooth Quick Complex
Waters Wedding Deep

TARDIS Morbius Kroll Power the Unquiet Pirate
Zygons Brain Dalek World Mutants Ice Ribos Light Destiny Ones
Invasion episode part Daleks Seeds Talons Monster Dinosaurs Warriors
Cyberman Terror Robot Spiders Assassin Mind Four Snowmen Eden Leisure
Part episode Devils Eyes Nature Lion Three Genesis Towers Deep

Carnival Deadly 10 13 Roof Army 42 Marinus Invisible Frontios Abominable Partners Holiday Opens Ways Decision 9 Trial 1 Rome

* For example, we count
14-episode story
The Trial of a Time Lord
(1986) 14 times.

151

## Time of the Daleks

Frequency of words used in Doctor Who story titles*

Enlightenment
Woman
Spaceship
Fireplace
Masque
Eleventh
Flatline
Delta
Ood
Listen
Rescue
Kill
Wardrobe
Sky
Rock
Doctors
Big
Hell
Boom
Pompeii
Sensorites
Zygon
Deep
Rings
Vengeance
Gridlock
Menace
Hide
Christmas
World
Green
Dalek
War
Patrol
Code
End
Unknown
Evolution
Love
Games
Robots
Tenth
City
New
Dark
Awakening
Earth
Age
River
Town
Dead
Ice
Hungry
Invasion
Plan
Blink
Power
Before
Sontaran
King's
Bannermen
Circle
Sun
Rise
Romans
Who
Stolen
Trial
Caves
Heaven
Stones
Ark
Dilemma
Makers
Marinus
State
Amy's
Dominators
Forest
Her
Savages
Blood
Wife
Marco
Bartholomews
Warriors'
Horns
Meglos
Doom
Destiny
Assassin
Robber
Pit
Marco
Celestial
Utopia
Empty
Fang
Cybermen
Midnight
Evil
Weng-Chiang
Three
Planet
Child
Unearthly
Time
Mind
Akhaten
Lord
Dæmons
Mars
Express
Full
a
Ghosts
Meddler
Nemesis
Experiment
Sound
Bells
Reunion
Paradise
Smith
Library
Seeds
Gunfighters
Androzani
Deadly
Terminus
Smugglers
Water
Venice
Aztecs
Tooth
Tara
Voyage
Bent
Machines
Demons
Reign
Peladon
Robot
Day
Astronaut
Factor
Mission
Hand
Edge
Almost
Satan
Flight
Giants
Hive
Chase
Ribos
Decay
Space
Asylum
Infinity
Aliens
Jones
on
Wasp
Happiness
Long
Undead
Inferno
Brain
Inversion
Androids
Doctor
Nightmare
Steel
Ambassadors
Snowmen
Damned
Curse
Colony
School
Unicorn
Earthshock
Daughter
Victory
Sleep
Spiders
Left
Fenric
Silence
Journey's
Man
Raven
Timelash
Lodger
Varos
Night
Vincent

* For example, we count
14-episode story
The Trial of a Time Lord
(1986) once as
it is a single story.

# Evolution of the Daleks

How the Doctor's deadliest enemies develop over the centuries!

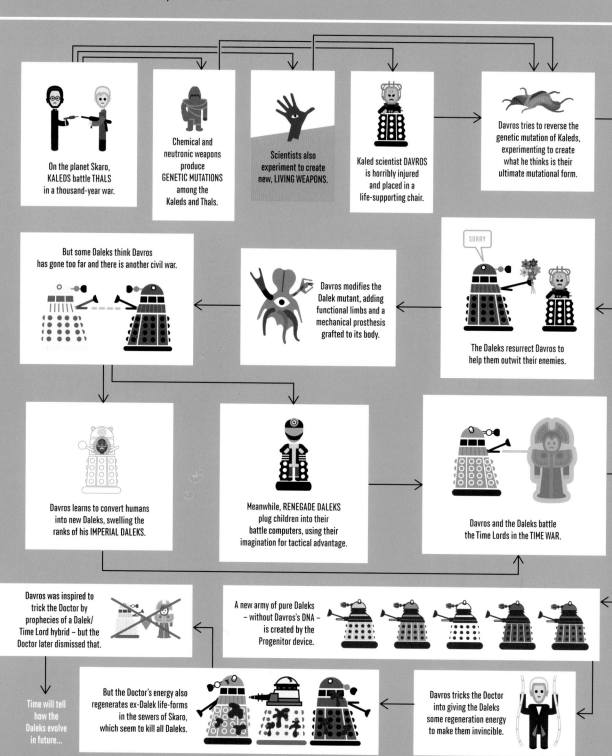

On the planet Skaro, KALEDS battle THALS in a thousand-year war.

Chemical and neutronic weapons produce GENETIC MUTATIONS among the Kaleds and Thals.

Scientists also experiment to create new, LIVING WEAPONS.

Kaled scientist DAVROS is horribly injured and placed in a life-supporting chair.

Davros tries to reverse the genetic mutation of Kaleds, experimenting to create what he thinks is their ultimate mutational form.

But some Daleks think Davros has gone too far and there is another civil war.

Davros modifies the Dalek mutant, adding functional limbs and a mechanical prosthesis grafted to its body.

SORRY

The Daleks resurrect Davros to help them outwit their enemies.

Davros learns to convert humans into new Daleks, swelling the ranks of his IMPERIAL DALEKS.

Meanwhile, RENEGADE DALEKS plug children into their battle computers, using their imagination for tactical advantage.

Davros and the Daleks battle the Time Lords in the TIME WAR.

Davros was inspired to trick the Doctor by prophecies of a Dalek/ Time Lord hybrid – but the Doctor later dismissed that.

A new army of pure Daleks – without Davros's DNA – is created by the Progenitor device.

Time will tell how the Daleks evolve in future...

But the Doctor's energy also regenerates ex-Dalek life-forms in the sewers of Skaro, which seem to kill all Daleks.

Davros tricks the Doctor into giving the Daleks some regeneration energy to make them invincible.

He places this creature inside a Mark III travel machine. These "Daleks" are conditioned to survive by dominating all other life forms.

Unfortunately, "all other life forms" includes Davros.

The Daleks leave Skaro, intent on becoming the supreme power in the universe.

The Doctor feeds some Daleks this HUMAN FACTOR, starting a civil war that almost wipes out the Daleks!

The Daleks test the Doctor's friend JAMIE to learn the HUMAN FACTOR that keeps defeating them.

Despite many conquests, the Daleks are continually defeated.

Dalek Caan travels back in time to rescue Davros.

In the first year of the Time War, Davros is lost when his command ship flies into the jaws of the NIGHTMARE CHILD at the GATES OF ELYSIUM.

Daleks now ruled by a DALEK EMPEROR.

After the Time War, a Dalek absorbs ROSE TYLER's DNA and begins to feel human emotions. It self-destructs.

The Dalek Emperor creates a new army of Daleks from dead humans – but all are destroyed by Rose Tyler/ the BAD WOLF.

Davros creates a new army of Daleks from his own flesh. DONNA NOBLE destroys them.

Dalek Sec creates a hybrid of himself and a human, which he thinks has a better chance of survival. Other Daleks exterminate him.

The Dalek Emperor creates the CULT OF SKARO – four Daleks with individual names (Sec, Thay, Caan and Jast), who can think like their enemies and devise new ways to survive.

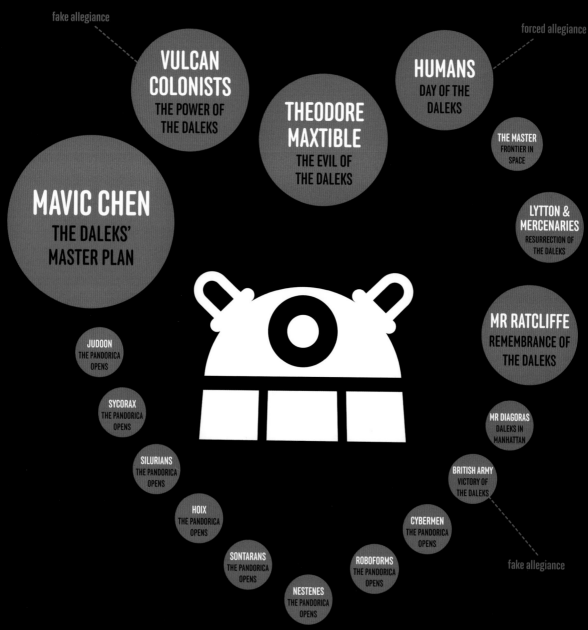

fake allegiance

forced allegiance

**VULCAN COLONISTS**
THE POWER OF THE DALEKS

**THEODORE MAXTIBLE**
THE EVIL OF THE DALEKS

**HUMANS**
DAY OF THE DALEKS

**THE MASTER**
FRONTIER IN SPACE

**MAVIC CHEN**
THE DALEKS' MASTER PLAN

**LYTTON & MERCENARIES**
RESURRECTION OF THE DALEKS

**MR RATCLIFFE**
REMEMBRANCE OF THE DALEKS

**JUDOON**
THE PANDORICA OPENS

**SYCORAX**
THE PANDORICA OPENS

**MR DIAGORAS**
DALEKS IN MANHATTAN

**SILURIANS**
THE PANDORICA OPENS

**BRITISH ARMY**
VICTORY OF THE DALEKS

**HOIX**
THE PANDORICA OPENS

**CYBERMEN**
THE PANDORICA OPENS

**SONTARANS**
THE PANDORICA OPENS

**ROBOFORMS**
THE PANDORICA OPENS

**NESTENES**
THE PANDORICA OPENS

fake allegiance

Sized according to the number
of episodes the allegiances last.

# ALLEGI

"You lot, working together! An alliance. How is that possible?"
→ The Eleventh Doctor, The Pandorica Opens (2010)

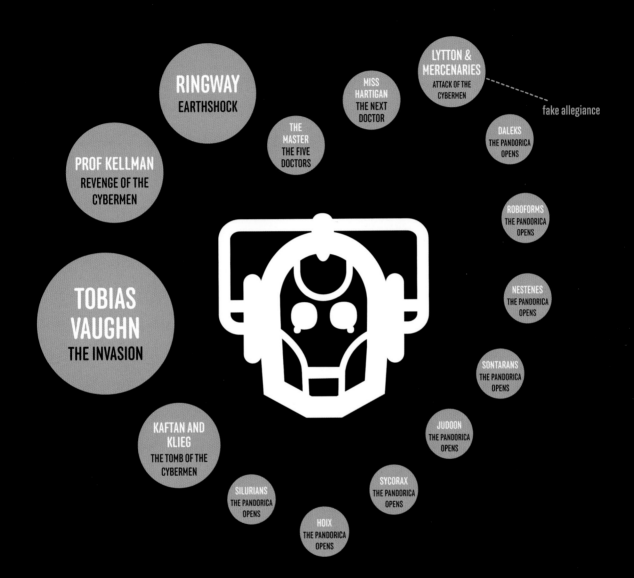

RINGWAY
**EARTHSHOCK**

MISS
HARTIGAN
THE NEXT
DOCTOR

LYTTON &
MERCENARIES
ATTACK OF THE
CYBERMEN

fake allegiance

THE
MASTER
THE FIVE
DOCTORS

DALEKS
THE PANDORICA
OPENS

PROF KELLMAN
**REVENGE OF THE
CYBERMEN**

ROBOFORMS
THE PANDORICA
OPENS

NESTENES
THE PANDORICA
OPENS

TOBIAS
VAUGHN
**THE INVASION**

SONTARANS
THE PANDORICA
OPENS

JUDOON
THE PANDORICA
OPENS

KAFTAN AND
KLIEG
THE TOMB OF THE
CYBERMEN

SYCORAX
THE PANDORICA
OPENS

SILURIANS
THE PANDORICA
OPENS

HOIX
THE PANDORICA
OPENS

THE PARTNERSHIPS OF THE DALEKS AND THE CYBERMEN

# ANCES

# THE TENTH DOCTOR
## VITAL STATISTICS

SONIC SCREWDRIVERS MARK V AND MARK VI

### TENTH DOCTOR TRANSLATIONS

"Molto Bene!"

"Very Well!" (ITALIAN)

# 47

## NUMBER OF (FULL-LENGTH) EPISODES

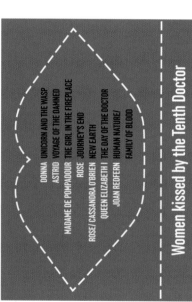

DONNA — UNICORN AND THE WASP
ASTRID — VOYAGE OF THE DAMNED
MADAME DE POMPADOUR — THE GIRL IN THE FIREPLACE
ROSE — JOURNEY'S END
NEW EARTH
ROSE/ CASSANDRA O'BRIEN
QUEEN ELIZABETH I — THE DAY OF THE DOCTOR
JOAN REDFERN — HUMAN NATURE/ FAMILY OF BLOOD

### Women kissed by the Tenth Doctor

"New teeth. That's weird."

### FIRST WORDS

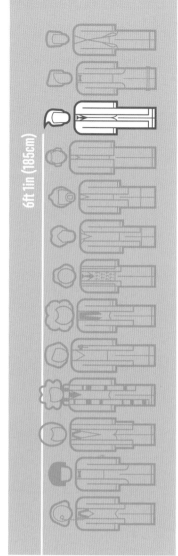

6ft 1in (185cm)

6
5
4
3
2
1

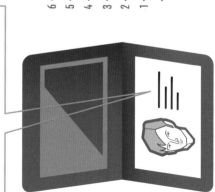

NEW EARTH 'WARD 26 – PLEASE COME' TOOTH AND CLAW DOCTORATE FROM THE UNIVERSITY OF EDINBURGH THE IDIOT'S LANTERN THE KING OF BELGIUM FEAR HER POLICE OFFICER THE SHAKESPEARE CODE SIR DOCTOR OF TARDIS WITH MISS MARTHA JONES VOYAGE OF THE DAMNED RED 6–7 PLUS ONE PARTNERS IN CRIME JOHN SMITH, HEALTH AND SAFETY FIRES OF POMPEII MARBLE INSPECTOR PLANET OF THE OOD THE DOCTOR AND DONNA NOBLE, REPRESENTING THE NOBLE CORPORATION PLC LIMITED INTERGALACTIC THE UNICORN AND THE WASP CHIEF INSPECTOR SMITH OF SCOTLAND YARD SILENCE IN THE LIBRARY 'THE LIBRARY. COME AS SOON AS YOU CAN x' MIDNIGHT ENGINE EXPERT, COMPANY INSURANCE PLANET OF THE DEAD OYSTER CARD (ALSO USED IN RISE OF THE CYBERMEN, ARMY OF GHOSTS AND EVOLUTION OF THE DALEKS)

→ PSYCHIC PAPER, THE TENNANT YEARS

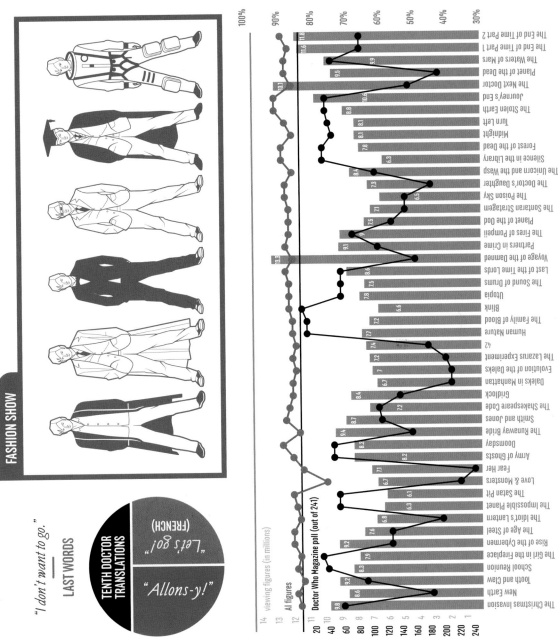

# InTime

The amount of time we see elapse in Doctor Who stories.

TIME ELAPSED

**4.5bn YEARS**

**1000 YEARS**

**14 YEARS**

**1 YEAR**

**3+ WEEKS**

**1 NIGHT**

**42 MINS**

**42**
(2007)

**THE HAPPINESS PATROL**
(1988)

**MARCO POLO**
(1964)

**THE POWER OF THREE**
(2012)

**THE ELEVENTH HOUR**
(2010)

**THE TIME OF THE DOCTOR**
(2013)

**HEAVEN SENT**
(2015)

NIGHT (2008) • MISSION TO THE UNKNOWN (1965) • DOCTOR WHO AND THE SILURIANS (1970) • THE MIND OF EVIL (1971) • THE DÆMONS (1971) • THE SEA DEVILS (1972) • THE SONTARAN EXPERIMENT (1975) • GENESIS OF THE DALEKS (1975) • MIDNIGHT

HEAVEN SENT (2015)

# Without the TARDIS

The TARDIS has been absent from 9 stories.

# Map of the TARDIS (2013)

The Doctor, Clara and the Van Baalen brothers are trapped
deep inside the TARDIS with monstrous versions of their own future selves...

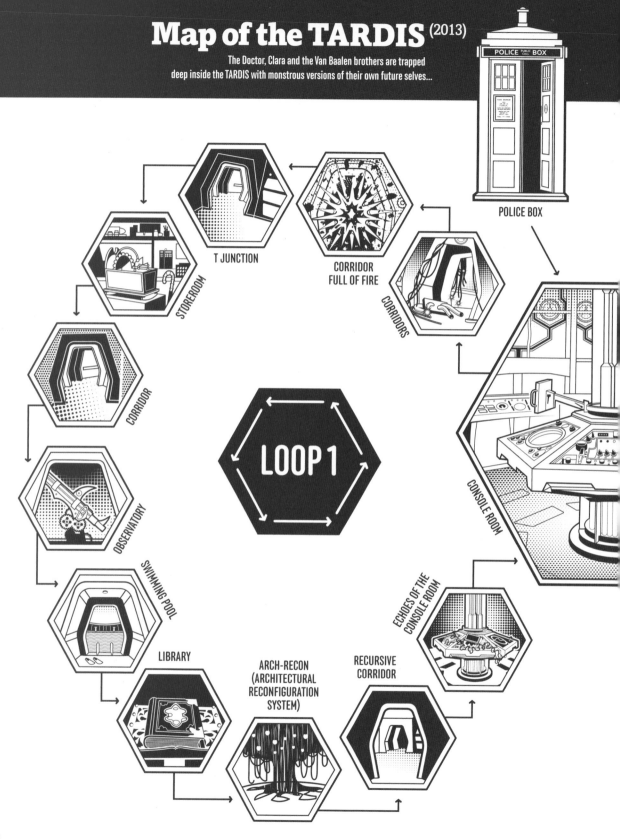

POLICE BOX

T JUNCTION

CORRIDOR
FULL OF FIRE

STOREROOM

CORRIDORS

CORRIDOR

CONSOLE ROOM

OBSERVATORY

LOOP 1

SWIMMING POOL

ECHOES OF THE
CONSOLE ROOM

LIBRARY

ARCH-RECON
(ARCHITECTURAL
RECONFIGURATION
SYSTEM)

RECURSIVE
CORRIDOR

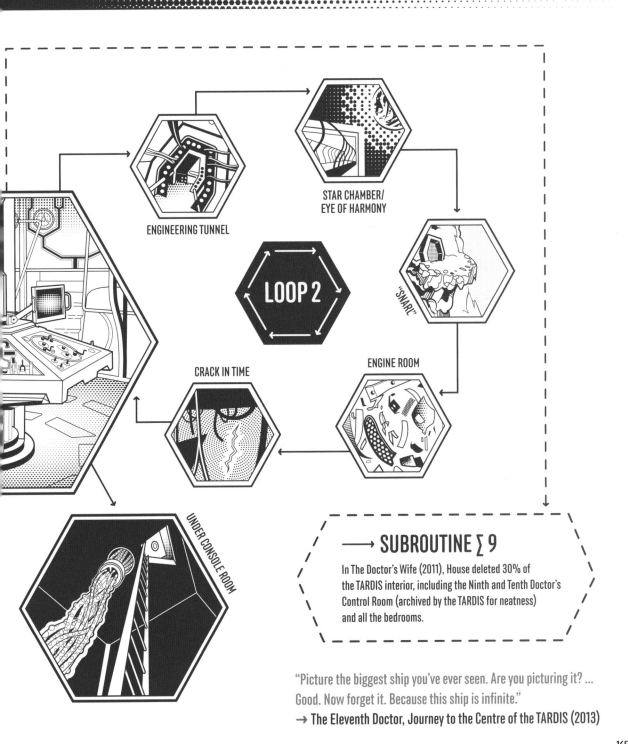

ENGINEERING TUNNEL

STAR CHAMBER/
EYE OF HARMONY

LOOP 2

"SNARL"

CRACK IN TIME

ENGINE ROOM

UNDER CONSOLE ROOM

## ⟶ SUBROUTINE ∑ 9

In The Doctor's Wife (2011), House deleted 30% of
the TARDIS interior, including the Ninth and Tenth Doctor's
Control Room (archived by the TARDIS for neatness)
and all the bedrooms.

"Picture the biggest ship you've ever seen. Are you picturing it? ...
Good. Now forget it. Because this ship is infinite."
→ The Eleventh Doctor, Journey to the Centre of the TARDIS (2013)

# Brown

V

All that running, you think he'd own more than two suits...

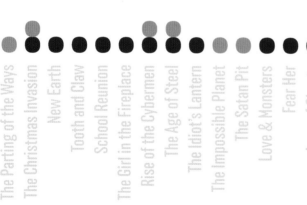

# S Blue

**BROWN 53%**
**BLUE 24%**
**OTHERS* 23%**

*(Leather Jacket,
Dressing Gown,
Dinner Suit,
Pyjamas
Spacesuit,
Schoolteacher
Uniform)

"You know, we have a very
extensive wardrobe here."
→ **The First Doctor,
The Edge of Destruction (1964)**

The Family of Blood
Blink
Utopia
The Sound of Drums
Last of the Time Lords
Voyage of the Damned
Partners in Crime
The Fires of Pompeii
Planet of the Ood
The Sontaran Stratagem
The Poison Sky
The Doctor's Daughter
The Unicorn and the Wasp
Silence in the Library
Forest of the Dead
Midnight
Turn Left
The Stolen Earth
Journey's End
The Next Doctor
Planet of the Dead
The Waters of Mars
The End of Time

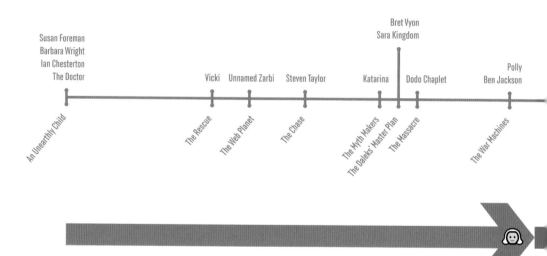

Susan Foreman
Barbara Wright
Ian Chesterton
The Doctor

An Unearthly Child

Vicki

The Rescue

Unnamed Zarbi

The Web Planet

Steven Taylor

The Chase

Katarina

The Myth Makers

Bret Vyon
Sara Kingdom

The Daleks' Master Plan

Dodo Chaplet

The Massacre

Polly
Ben Jackson

The War Machines

# EVERYONE WE'VE

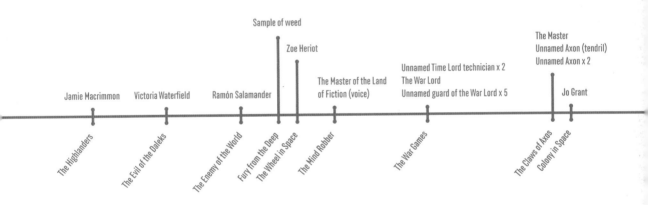

Jamie Macrimmon

Victoria Waterfield

Ramón Salamander

Sample of weed

Zoe Heriot

The Master of the Land
of Fiction (voice)

Unnamed Time Lord technician x 2
The War Lord
Unnamed guard of the War Lord x 5

The Master
Unnamed Axon (tendril)
Unnamed Axon x 2

Jo Grant

The Highlanders

The Evil of the Daleks

The Enemy of the World

Fury from the Deep

The Wheel in Space

The Mind Robber

The War Games

The Claws of Axos

Colony in Space

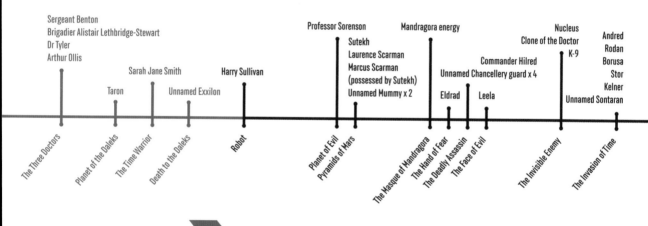

Sergeant Benton
Brigadier Alistair Lethbridge-Stewart
Dr Tyler
Arthur Ollis

Professor Sorenson

Mandragora energy

Nucleus
Clone of the Doctor

Andred
Rodan
Borusa
Stor
Kelner
Unnamed Sontaran

K-9

Sutekh
Laurence Scarman
Marcus Scarman
(possessed by Sutekh)
Unnamed Mummy x 2

Sarah Jane Smith

Harry Sullivan

Commander Hilred
Unnamed Chancellery guard x 4

Taron

Unnamed Exxilon

Eldrad

Leela

The Three Doctors

Planet of the Daleks

The Time Warrior

Death to the Daleks

Robot

Planet of Evil
Pyramids of Mars

The Masque of Mandragora

The Hand of Fear

The Deadly Assassin

The Face of Evil

The Invisible Enemy

The Invasion of Time

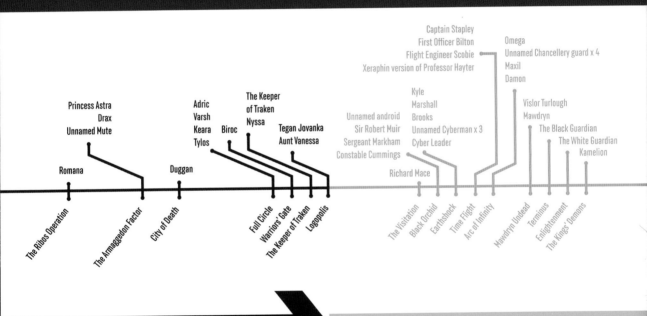

Princess Astra
Drax
Unnamed Mute

Romana

The Ribos Operation

The Armageddon Factor

City of Death

Duggan

Adric
Varsh
Keara
Tylos

Biroc

The Keeper
of Traken
Nyssa

Tegan Jovanka
Aunt Vanessa

Full Circle
Warriors' Gate
The Keeper of Traken
Logopolis

Captain Stapley
First Officer Bilton
Flight Engineer Scobie
Xeraphin version of Professor Hayter

Omega
Unnamed Chancellery guard x 4
Maxil
Damon

Kyle
Marshall
Brooks

Unnamed android
Sir Robert Muir
Sergeant Markham
Constable Cummings

Unnamed Cyberman x 3
Cyber Leader

Richard Mace

Vislor Turlough
Mawdryn

The Black Guardian
The White Guardian
Kamelion

The Visitation
Black Orchid
Earthshock
Time Flight
Arc of Infinity
Mawdryn Undead
Terminus
Enlightenment
The Kings' Demons

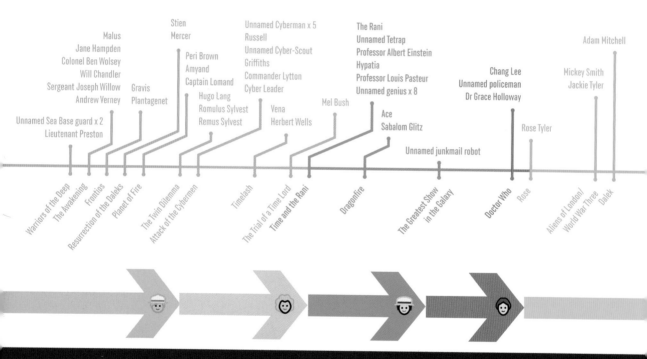

Malus
Jane Hampden
Colonel Ben Wolsey
Will Chandler
Sergeant Joseph Willow
Andrew Verney

Stien
Mercer

Peri Brown
Amyand
Captain Lomand

Unnamed Cyberman x 5
Russell
Unnamed Cyber-Scout
Griffiths
Commander Lytton
Cyber Leader

The Rani
Unnamed Tetrap
Professor Albert Einstein
Hypatia
Professor Louis Pasteur
Unnamed genius x 8

Adam Mitchell

Chang Lee
Unnamed policeman
Dr Grace Holloway

Mickey Smith
Jackie Tyler

Unnamed Sea Base guard x 2
Lieutenant Preston

Gravis
Plantagenet

Hugo Lang
Romulus Sylvest
Remus Sylvest

Vena
Herbert Wells

Mel Bush

Ace
Sabalom Glitz

Rose Tyler

Unnamed junkmail robot

Warriors of the Deep
The Awakening
Frontios
Resurrection of the Daleks
Planet of Fire
The Twin Dilemma
Attack of the Cybermen
Timelash
The Trial of a Time Lord
Time and the Rani
Dragonfire
The Greatest Show in the Galaxy
Doctor Who
Rose
Aliens of London/ World War Three
Dalek

172

# INSIDE THE

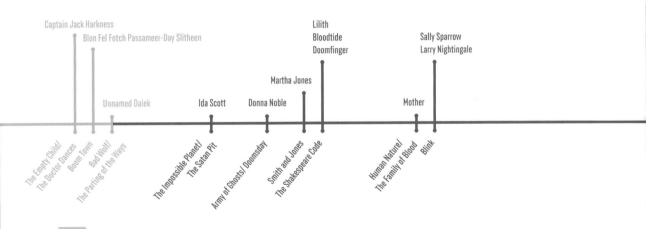

Captain Jack Harkness

Blon Fel Fotch Passameer-Day Slitheen

Lilith
Bloodtide
Doomfinger

Sally Sparrow
Larry Nightingale

Unnamed Dalek

Ida Scott

Donna Noble

Martha Jones

Mother

The Empty Child/
The Doctor Dances

Boom Town

Bad Wolf/
The Parting of the Ways

The Impossible Planet/
The Satan Pit

Army of Ghosts/ Doomsday

Smith and Jones

The Shakespeare Code

Human Nature/
The Family of Blood

Blink

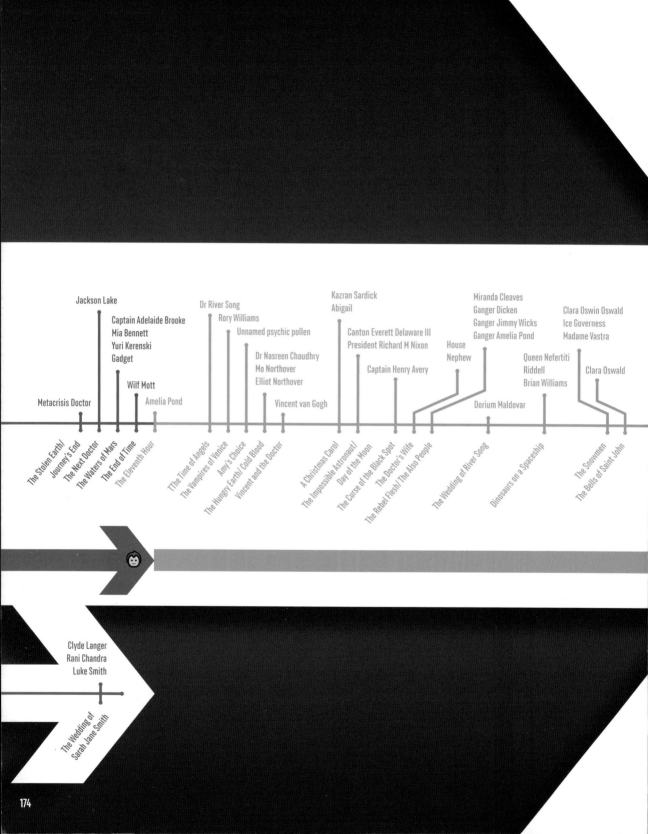

Metacrisis Doctor

Jackson Lake

Captain Adelaide Brooke
Mia Bennett
Yuri Kerenski
Gadget

Wilf Mott

Amelia Pond

Dr River Song
Rory Williams

Unnamed psychic pollen

Dr Nasreen Chaudhry
Mo Northover
Elliot Northover

Vincent van Gogh

Kazran Sardick
Abigail

Canton Everett Delaware III
President Richard M Nixon

Captain Henry Avery

House
Nephew

Dorium Maldovar

Miranda Cleaves
Ganger Dicken
Ganger Jimmy Wicks
Ganger Amelia Pond

Queen Nefertiti
Riddell
Brian Williams

Clara Oswin Oswald
Ice Governess
Madame Vastra

Clara Oswald

The Stolen Earth/
Journey's End
The Next Doctor
The Waters of Mars
The End of Time
The Eleventh Hour

TThe Time of Angels
The Vampires of Venice
Amy's Choice
The Hungry Earth/ Cold Blood
Vincent and the Doctor

A Christmas Carol
The Impossible Astronaut/
Day of the Moon
The Curse of the Black Spot
The Doctor's Wife
The Rebel Flesh/ The Also People

The Wedding of River Song

Dinosaurs on a Spaceship

The Snowmen
The Bells of Saint John

Clyde Langer
Rani Chandra
Luke Smith

The Wedding of
Sarah Jane Smith

174

# TARDIS*

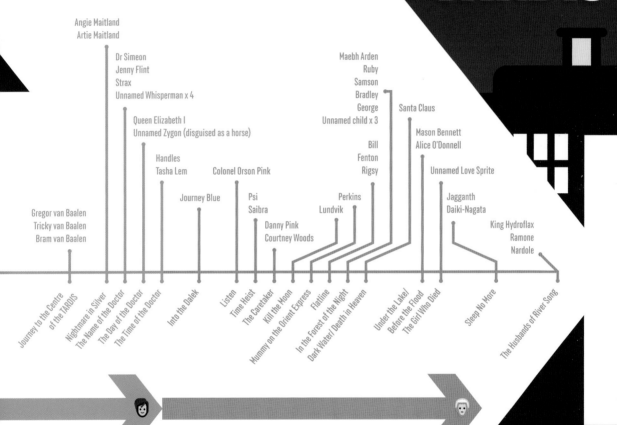

Angie Maitland
Artie Maitland

Dr Simeon
Jenny Flint
Strax
Unnamed Whisperman x 4

Maebh Arden
Ruby
Samson
Bradley
George
Unnamed child x 3

Santa Claus

Queen Elizabeth I
Unnamed Zygon (disguised as a horse)

Mason Bennett
Alice O'Donnell

Bill
Fenton
Rigsy

Handles
Tasha Lem

Unnamed Love Sprite

Colonel Orson Pink

Jagganth
Daiki-Nagata

Journey Blue

Psi
Saibra

Perkins
Lundvik

King Hydroflax
Ramone
Nardole

Gregor van Baalen
Tricky van Baalen
Bram van Baalen

Danny Pink
Courtney Woods

Journey to the Centre of the TARDIS
Nightmare in Silver
The Name of the Doctor
The Day of the Doctor
The Time of the Doctor
Into the Dalek
Listen
Time Heist
The Caretaker
Kill the Moon
Mummy on the Orient Express
Flatline
In the Forest of the Night
Dark Water/ Death in Heaven
Under the Lake/ Before the Flood
The Girl Who Died
Sleep No More
The Husbands of River Song

*We've included everyone we see:
1. Inside the TARDIS control room, corridors or other interior spaces.
2. Crossing the threshold of the police box exterior.

175

This chapter is 7% of Whographica.

# THE ELEVENTH DOCTOR
## VITAL STATISTICS

*"Legs!*
*I've still got legs!*
*Good!"*

——— FIRST WORDS

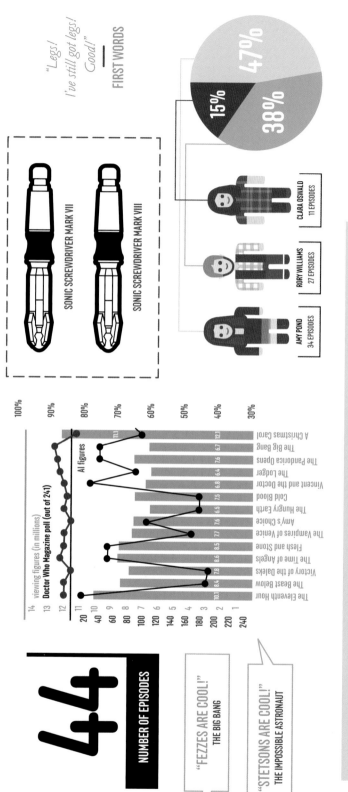

47%

38%

15%

SONIC SCREWDRIVER MARK VII

SONIC SCREWDRIVER MARK VIII

**AMY POND**
34 EPISODES

**RORY WILLIAMS**
27 EPISODES

**CLARA OSWALD**
11 EPISODES

**"BOW TIES ARE COOL!"**
THE ELEVENTH HOUR
AMY'S CHOICE
VINCENT AND THE DOCTOR
THE LODGER
THE SNOWMEN

viewing figures (in millions)

Doctor Who Magazine poll (out of 241)

AI figures

100%
90%
80%
70%
60%
50%
40%
30%

The Eleventh Hour — 10.1
The Beast Below — 8.4
Victory of the Daleks — 7.8
The Time of Angels — 8.6
Flesh and Stone — 8.5
The Vampires of Venice — 7.7
Amy's Choice — 7.6
The Hungry Earth — 6.5
Cold Blood — 7.5
Vincent and the Doctor — 6.8
The Lodger — 6.4
The Pandorica Opens — 7.6
The Big Bang — 6.7
A Christmas Carol — 12.1

11.1

14
13
12

11
10
9
8
7
6
5
4
3
2
1

20
40
60
80
100
120
140
160
180
200
220
240

**44**
NUMBER OF EPISODES

**"FEZZES ARE COOL!"**
THE BIG BANG

**"STETSONS ARE COOL!"**
THE IMPOSSIBLE ASTRONAUT

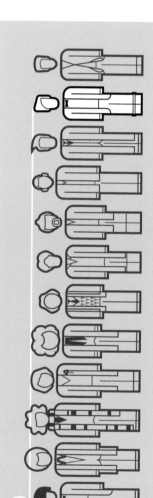

5ft 11in (180cm)

6
5
4
3
2
1

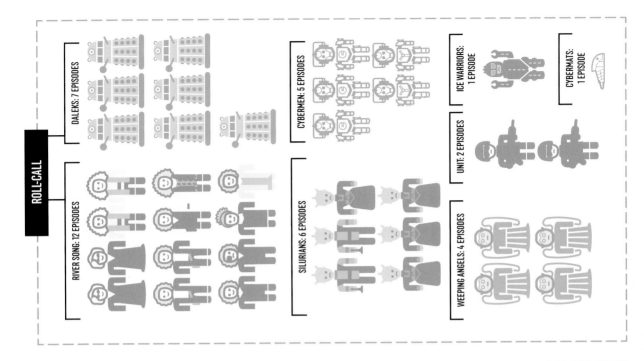

DALEKS: 7 EPISODES

RIVER SONG: 12 EPISODES

CYBERMEN: 5 EPISODES

SILURIANS: 6 EPISODES

ICE WARRIORS: 1 EPISODE

CYBERMATS: 1 EPISODE

UNIT: 2 EPISODES

WEEPING ANGELS: 4 EPISODES

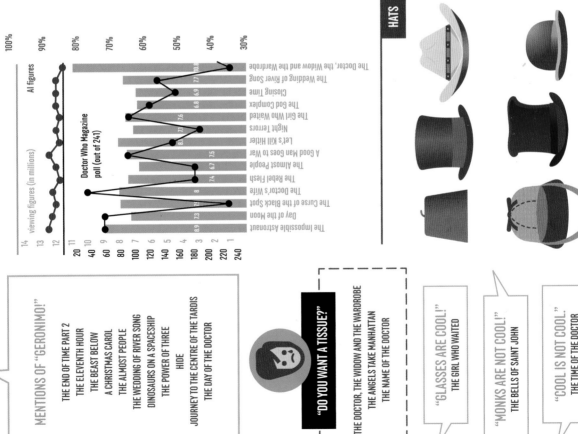

## MENTIONS OF "GERONIMO!"

THE END OF TIME PART 2
THE ELEVENTH HOUR
THE BEAST BELOW
A CHRISTMAS CAROL
THE ALMOST PEOPLE
THE WEDDING OF RIVER SONG
DINOSAURS ON A SPACESHIP
THE POWER OF THREE
HIDE
JOURNEY TO THE CENTRE OF THE TARDIS
THE DAY OF THE DOCTOR

### "DO YOU WANT A TISSUE?"

THE DOCTOR, THE WIDOW AND THE WARDROBE
THE ANGELS TAKE MANHATTAN
THE NAME OF THE DOCTOR

"GLASSES ARE COOL!"
THE GIRL WHO WAITED

"MONKS ARE NOT COOL!"
THE BELLS OF SAINT JOHN

"COOL IS NOT COOL."
THE TIME OF THE DOCTOR

### HATS

**Chart labels:**

AI figures

viewing figures (in millions)

Doctor Who Magazine poll (out of 241)

The Impossible Astronaut — 8.9 7.3 74
Day of the Moon
The Curse of the Black Spot
The Doctor's Wife — 7.4 8
The Rebel Flesh — 6.7 7.5
The Almost People
A Good Man Goes to War — 8.1 7.6
Let's Kill Hitler
Night Terrors — 7.1 7.6
The Girl Who Waited
The God Complex — 6.8 6.9
Closing Time
The Wedding of River Song — 7.7 10.8
The Doctor, the Widow and the Widow and the Wardrobe

100%
90%
80%
70%
60%
50%
40%
30%

14 13 12

20 40 60 80 100 120 140 160 180 200 220 240

11 10 9 8 7 6 5 4 3 2 1

# THE ELEVENTH DOCTOR

## VITAL STATISTICS

FASHION SHOW

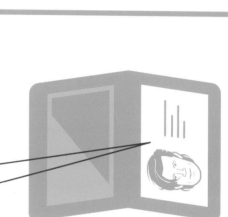

THE ELEVENTH HOUR TELEVISION REPAIR MAN
THE VAMPIRES OF VENICE CHURCH CREDENTIALS
THE HUNGRY EARTH MINISTRY OF DRILLS, EARTH AND
SCIENCE VINCENT AND THE DOCTOR MINISTRY OF ART
AND ARTYNESS THE LODGER THE DOCTOR'S
NATIONAL INSURANCE NUMBER, NHS NUMBER
AND REFERENCE FROM THE ARCHBISHOP OF
CANTERBURY THE REBEL FLESH METEOROLOGICAL
DEPARTMENT NIGHT TERRORS SOCIAL SERVICES
DEPARTMENT THE ANGELS TAKE MANHATTAN SPECIAL
COMMISSIONER FROM THE CHINESE EMPEROR HIDE
MILITARY INTELLIGENCE NIGHTMARE IN SILVER
PROCONSUL → PSYCHIC PAPER, THE SMITH YEARS

*"I will not forget one line of this.*
*Not one day. I swear.*
*I will always remember when*
*the Doctor was me."*

LAST WORDS

"BUNK BEDS ARE COOL!"
THE DOCTOR'S WIFE

viewing figures (in millions)    AI figures

Doctor Who Magazine poll (out of 241)

| Episode | Values |
|---|---|
| Asylum of the Daleks | 8.3 7.6 |
| Dinosaurs on a Spaceship | 7.6 8.4 |
| A Town Called Mercy | 8.4 7.7 |
| The Power of Three | 7.8 |
| The Angels Take Manhattan | 7.8 9.9 |
| The Snowmen | 9.9 8.4 |
| The Bells of Saint John | 8.4 7.5 |
| The Rings of Akhaten | 7.5 7.4 |
| Cold War | 7.4 6.6 |
| Hide | 6.6 |
| Journey to the Centre of the TARDIS | 6.5 6.5 |
| The Crimson Horror | 6.5 6.6 |
| Nightmare in Silver | 7.5 |
| The Name of the Doctor | 7.5 |
| The Day of the Doctor | 12.8 11.1 |
| The Time of the Doctor | 11.1 |

20/11  40/10  60/9  80/8  100/7  120/6  140/5  160/4  180/3  200/2  220/1  240

30%  40%  50%  60%  70%  80%  90%  100%

# OM NOM NOM

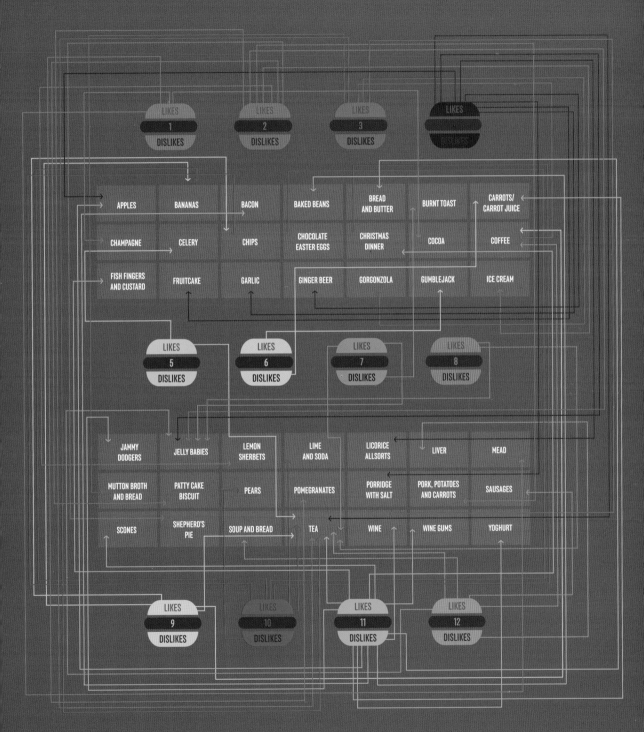

"Why can't you give me any decent food? You're Scottish. Fry something." → The Eleventh Doctor, The Eleventh Hour (2010)

181

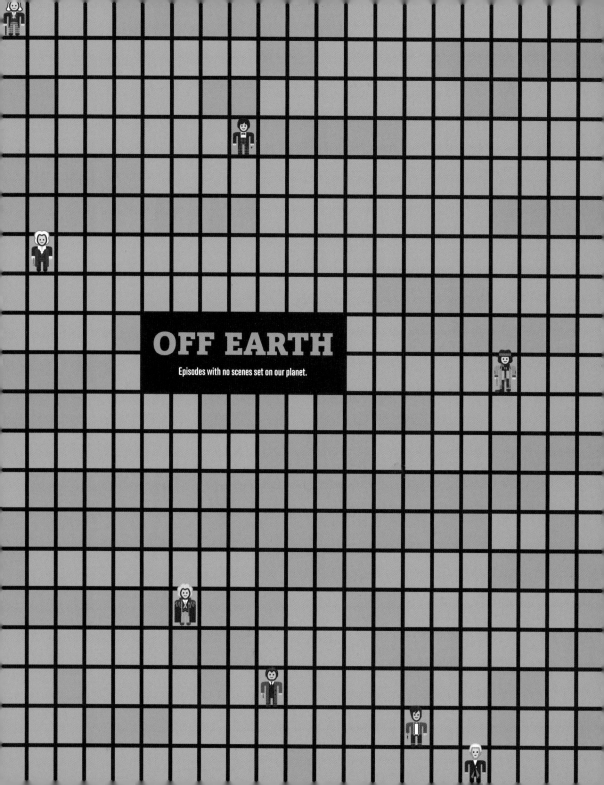

# OFF EARTH

Episodes with no scenes set on our planet.

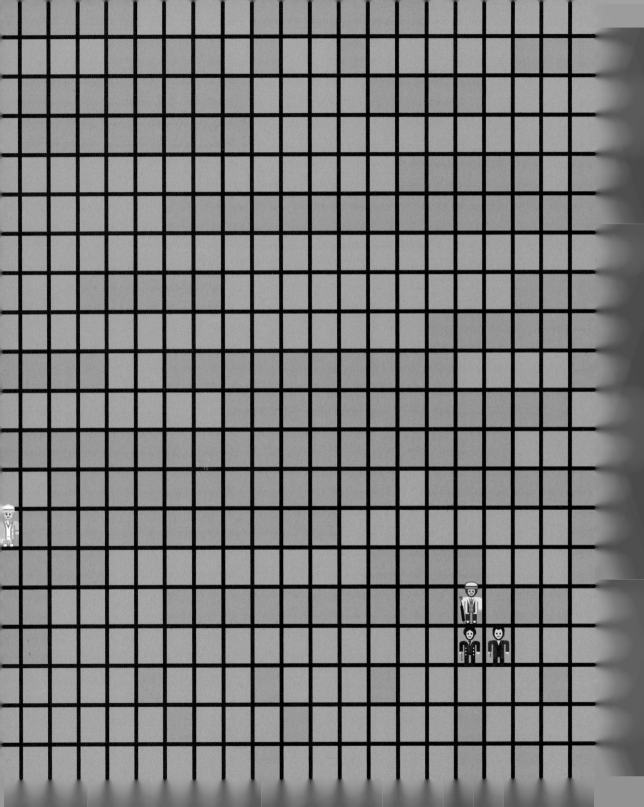

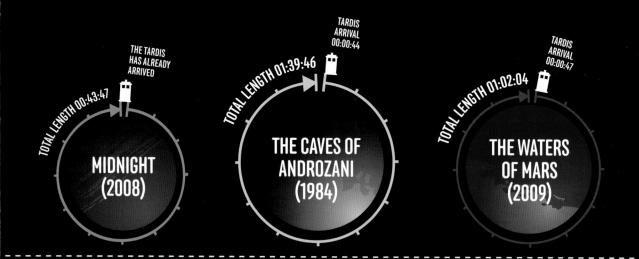

TOTAL LENGTH 00:43:47

THE TARDIS
HAS ALREADY
ARRIVED

**MIDNIGHT
(2008)**

TOTAL LENGTH 01:39:46

TARDIS
ARRIVAL
00:00:44

**THE CAVES OF
ANDROZANI
(1984)**

TOTAL LENGTH 01:02:04

TARDIS
ARRIVAL
00:00:47

**THE WATERS
OF MARS
(2009)**

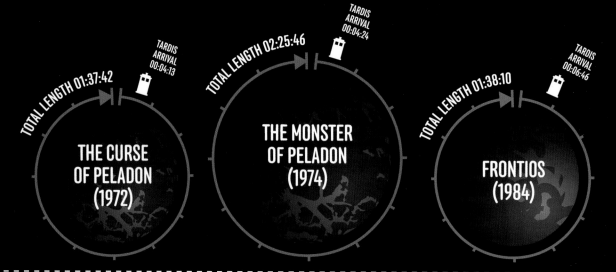

TOTAL LENGTH 01:37:42

TARDIS
ARRIVAL
00:04:13

**THE CURSE
OF PELADON
(1972)**

TOTAL LENGTH 02:25:46

TARDIS
ARRIVAL
00:04:24

**THE MONSTER
OF PELADON
(1974)**

TOTAL LENGTH 01:38:10

TARDIS
ARRIVAL
00:06:46

**FRONTIOS
(1984)**

# Planetary
# Arrivals

Alien worlds named in titles of Doctor Who stories
and how long it takes the TARDIS to land there in those stories.

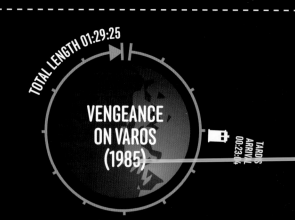

TOTAL LENGTH 01:29:25

**VENGEANCE
ON VAROS
(1985)**

TARDIS
ARRIVAL
00:23:04

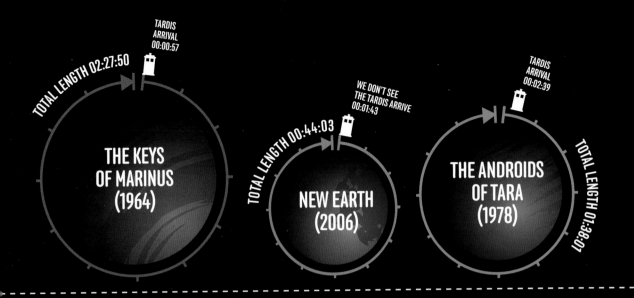

TARDIS
ARRIVAL
00:00:57

TOTAL LENGTH 02:27:50

THE KEYS
OF MARINUS
(1964)

WE DON'T SEE
THE TARDIS ARRIVE
00:01:43

TOTAL LENGTH 00:44:03

NEW EARTH
(2006)

TARDIS
ARRIVAL
00:02:39

TOTAL LENGTH 01:38:07

THE ANDROIDS
OF TARA
(1978)

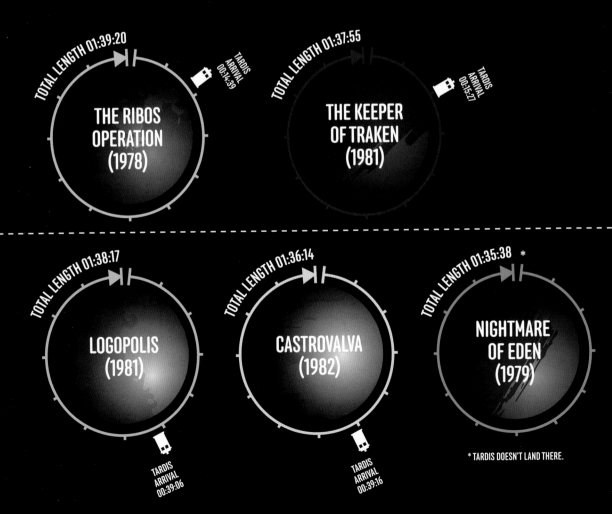

TOTAL LENGTH 01:39:20

THE RIBOS
OPERATION
(1978)

TARDIS
ARRIVAL
00:14:39

TOTAL LENGTH 01:37:55

THE KEEPER
OF TRAKEN
(1981)

TARDIS
ARRIVAL
00:35:27

TOTAL LENGTH 01:38:17

LOGOPOLIS
(1981)

TARDIS
ARRIVAL
00:39:06

TOTAL LENGTH 01:36:14

CASTROVALVA
(1982)

TARDIS
ARRIVAL
00:39:16

TOTAL LENGTH 01:35:38 *

NIGHTMARE
OF EDEN
(1979)

* TARDIS DOESN'T LAND THERE.

185

# Meet the Neighbours

Our Solar System, according to Doctor Who.

1 ● MERCURY

2 ● VENUS

Has metal seas (Marco Polo (1964)), flowers (The Wheel in Space (1968)), spearmint
(The Shakespeare Code (2007)) and a creature called a Shanghorn (The Green Death (1973)).
The Doctor is skilled in Venusian aikido, first seen in Inferno (1970).

3 ● EARTH

The Doctor's favourite planet (see pages 56–57).

4 ● MARS

The Doctor visits in The Waters of Mars (2009). Indigenous species include the Ice Warriors
and the Flood. The Osirans also built at least one pyramid on Mars in the time of
the ancient Egyptians (Pyramids of Mars (1975)).

5 ● UNNAMED FIFTH PLANET

Home of the alien Fendahl destroyed by the Time Lords 12 million years ago,
possibly creating the asteroid belt (Image of the Fendahl (1977)).

6 ● JUPITER

British astronaut Guy Crayford was rescued by the Kraals while trapped in orbit around Jupiter
(The Android Invasion (1975)). In the future, the 'planet' of gold, Neo Phobos, moves into orbit
around Jupiter as one of its moons and is renamed 'Voga' (Revenge of the Cybermen (1975)).

7 ● SATURN

The Doctor is infected by a talking virus while on Saturn's moon, Titan (The Invisible Enemy (1977)).
It seems Titan is destroyed at the end of the story.

8 ● URANUS

The only source of the rare mineral taranium, used to build a time destructor (The Daleks' Master Plan (1965–6)).

9 ● NEPTUNE

In the 38th century, the Le Verrier space station is in orbit round Neptune and there are
colonies on the planet's largest moon, Triton (Sleep No More (2015)).

10 ● MONDAS

Home of the Cybermen and originally twin of Earth, Mondas drifted out of the Solar System,
returning in 1986 – where it was destroyed (The Tenth Planet (1966)).

11 ● PLUTO

The Fourth Doctor calls it a planet when he visits in The Sun Makers (1977).
One of its inhabitants, Mandrell, doesn't believe there is life on any other planets.

12 ● CASSIUS

According to K-9 in The Sun Makers, Pluto was thought to be the 'outermost body'
of the Solar System until the discovery of this world.

13 ● UNKNOWN PLANET

14 ● PLANET 14

Used as a base by the Cybermen in The Invasion (1968) – though it's not stated
as such that it is in the Solar System.

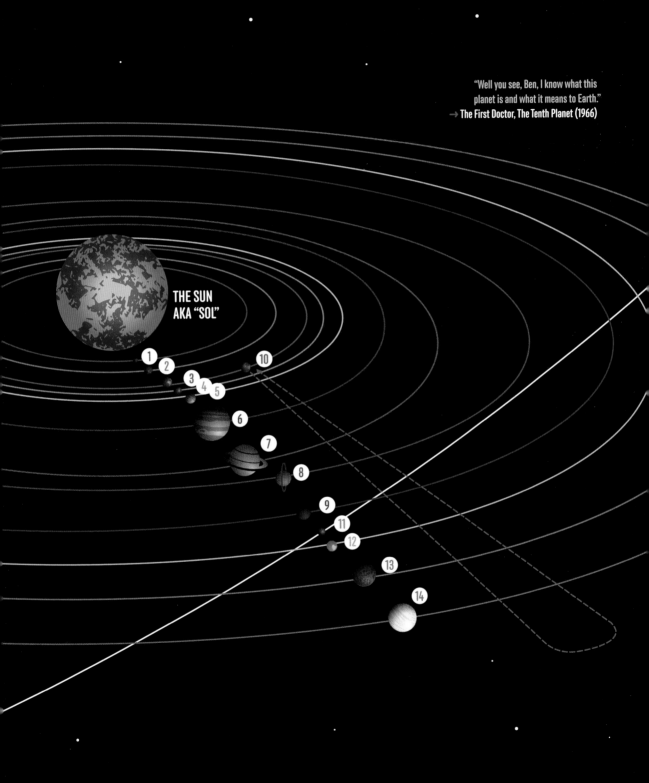

"Well you see, Ben, I know what this
planet is and what it means to Earth."
→ The First Doctor, The Tenth Planet (1966)

THE SUN
AKA "SOL"

# WHERE IS GALLIFREY?

Using clues from the series to track down the Doctor's missing home planet.

"Gallifrey is currently positioned at the extreme end of the time continuum, for its own protection. We're at the end of the universe, give or take a star system."
→ The General, Hell Bent (2015)

In Utopia, the Tenth Doctor says the end of the universe is sometime beyond the year 100,000,000,000,000.

That suggests Gallifrey is visible from Earth with the naked eye – so within about 16,000 light years. But Susan might not mean it literally and is just saying home is very distant.

The Time Lord who comes to Earth at the start of Terror of the Autons (1971) claims to have travelled 29,000 light years. If he's come from Gallifrey, that puts it in or close to our galaxy, the Milky Way, which is thought to be 100,000 light years across. Earth is about halfway between the outer rim and the central core.

We're told Gallifrey is much older than Earth. The orange sunlight on Gallifrey suggests its main sun (it has two) is very old.

We now think the core is a bit closer, so 29,000 light years means Gallifrey is just on the far side of the galactic core from us.

Gallifrey might be too distant from Earth – in space as well as time – for us to see it, even with our best telescopes. But if our deductions are right, we at least know in which direction to look. The galactic core is in the constellation of Sagittarius at co-ordinates RA: 17h 45m 40.04s DEC: 29° 00' 28.1".

## WHERE

In Pyramids of Mars (1975), the Doctor gives the binary co-ordinates of Gallifrey from galactic zero centre as 1001100x02. But that don't help us locate Gallifrey because:

**1** We don't know where "galactic zero centre" is – it might be the centre of this or another galaxy, or an accepted but arbitrary intergalactic reference point.

**2** We don't know the scale being used, which makes a big difference. For example, Earth astronomers use different scales: kilometres; astronomical units (1 AU = about 150 million km); light years (1 LY = about 9.5 trillion km); and parsecs (1 pc = about 31 trillion km).

**3** In mathematics, a binary number should be composed of only two digits – usually 0 and 1. The co-ordinate "02" suggests Gallifrey's location is unusual, perhaps reaching into another dimension.

What other clues has the series given us? The first reference to the location of the Doctor's home planet is in Marco Polo (1964). His granddaughter Susan says her home is "as far away as a night star".

So perhaps it is located in one of the globular clusters of stars "above" or "below" the galactic disc, which astronomers now think could contain ancient, habitable planets.

On the disc itself, older stars are nearer the core. When Terror of the Autons was made, we thought the distance from Earth to the galactic core was 29,000 light years – the figure cited by the Time Lord – so it seems that's where the people making Doctor Who intended it to be.

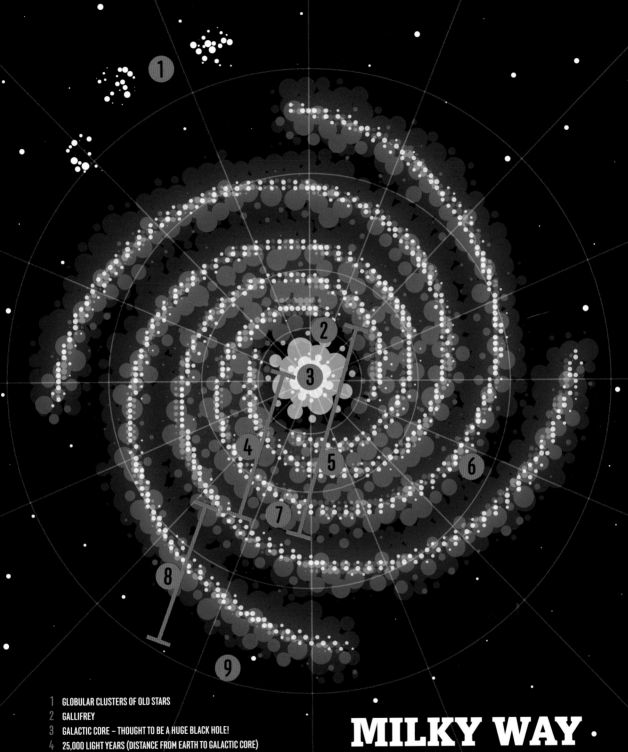

1 GLOBULAR CLUSTERS OF OLD STARS
2 GALLIFREY
3 GALACTIC CORE – THOUGHT TO BE A HUGE BLACK HOLE!
4 25,000 LIGHT YEARS (DISTANCE FROM EARTH TO GALACTIC CORE)
5 29,000 LIGHT YEARS (DISTANCE FROM EARTH TO GALLIFREY)
6 WHAT THE TIME LORDS CALL "MUTTER'S SPIRAL"
7 EARTH
8 25,000 LIGHT YEARS (DISTANCE FROM EARTH TO OUTER RIM)
9 OUTER RIM

# MILKY WAY
## GALAXY*

*Known by the Time Lords as the "Stellian" galaxy

# THE TWELFTH DOCTOR:
## VITAL STATISTICS

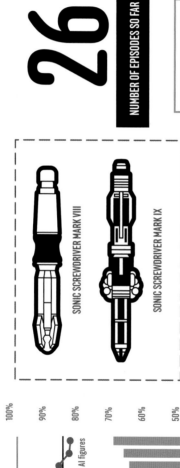

### MENTIONS OF "SHUT UP!"

DEEP BREATH
LISTEN
TIME HEIST
THE CARETAKER
DARK WATER
LAST CHRISTMAS
FACE THE RAVEN

## 26

### NUMBER OF EPISODES SO FAR

### WORST INSULTS

"Execute me now. Hang on, execute him.
I'd like to see if his head keeps laughing
when you chop it off."
**TO ROBIN HOOD,
ROBOT OF SHERWOOD**

"Why can't I meet a decent species?
Planet of the pudding brains!"
**DEEP BREATH**

"Listen, when we're done here,
by all means go and find yourself
a shoulder to cry on. You'll
probably need that. Until then,
what you need is me!" **LISTEN**

SONIC SCREWDRIVER MARK VIII

SONIC SCREWDRIVER MARK IX

*"Kidneys!
I've got new kidneys!"*

### LAST WORDS

6ft (183cm)

6
5
4
3
2
1

### viewing figures (in millions)

100%
90%
80%
70%
60%
50%
40%
30%

All figures

Doctor Who Magazine poll –
no data as of publication date

| | |
|---|---|
| Last Christmas | 8.3 |
| Death in Heaven | 7.6 |
| Dark Water | 7.3 |
| In the Forest of the Night | 6.9 |
| Flatline | 6.7 |
| Mummy on the Orient Express | 7.1 |
| Kill the Moon | 6.9 |
| The Caretaker | 6.8 |
| Time Heist | 7 |
| Listen | 7 |
| Robot of Sherwood | 7.3 |
| Into the Dalek | 7.3 |
| Deep Breath | 9.2 |

14
13
12

20
40
60
80
100
120
140
160
180
200
220
240

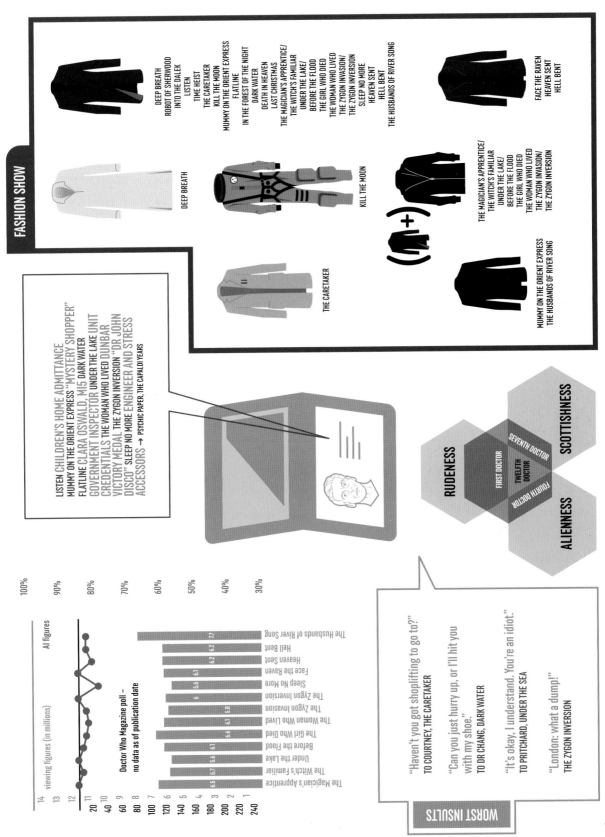

**FASHION SHOW**

DEEP BREATH
ROBOT OF SHERWOOD
INTO THE DALEK
LISTEN
TIME HEIST
THE CARETAKER
KILL THE MOON
MUMMY ON THE ORIENT EXPRESS
FLATLINE
IN THE FOREST OF THE NIGHT
DARK WATER
DEATH IN HEAVEN
LAST CHRISTMAS
THE MAGICIAN'S APPRENTICE/
THE WITCH'S FAMILIAR
UNDER THE LAKE/
BEFORE THE FLOOD
THE GIRL WHO DIED
THE WOMAN WHO LIVED
THE ZYGON INVASION/
THE ZYGON INVERSION
SLEEP NO MORE
HEAVEN SENT
HELL BENT
THE HUSBANDS OF RIVER SONG

FACE THE RAVEN
HEAVEN SENT
HELL BENT

DEEP BREATH

KILL THE MOON

THE CARETAKER

THE MAGICIAN'S APPRENTICE/
THE WITCH'S FAMILIAR
UNDER THE LAKE/
BEFORE THE FLOOD
THE GIRL WHO DIED
THE WOMAN WHO LIVED
THE ZYGON INVASION/
THE ZYGON INVERSION

MUMMY ON THE ORIENT EXPRESS
THE HUSBANDS OF RIVER SONG

LISTEN CHILDREN'S HOME ADMITTANCE
MUMMY ON THE ORIENT EXPRESS "MYSTERY SHOPPER"
FLATLINE CLARA OSWALD, MI5 DARK WATER
GOVERNMENT INSPECTOR UNDER THE LAKE UNIT
CREDENTIALS THE WOMAN WHO LIVED DUNBAR
VICTORY MEDAL THE ZYGON INVERSION "DR JOHN
DISCO" SLEEP NO MORE ENGINEER AND STRESS
ACCESSORS → PSYCHIC PAPER, THE CAPALDI YEARS

RUDENESS

SCOTTISHNESS

SEVENTH DOCTOR

FIRST DOCTOR

TWELFTH DOCTOR

FOURTH DOCTOR

ALIENNESS

**WORST INSULTS**

"Haven't you got shoplifting to go to?"
TO COURTNEY, THE CARETAKER

"Can you just hurry up, or I'll hit you
with my shoe."
TO DR CHANG, DARK WATER

"It's okay, I understand. You're an idiot."
TO PRITCHARD, UNDER THE SEA

"London: what a dump!"
THE ZYGON INVERSION

viewing figures (in millions)    All figures

14
13
12
11
10
9
8

100%
90%
80%
70%
60%
50%
40%
30%

Doctor Who Magazine poll –
no data as of publication date

240
220
200
180
160
140
120
100
80
60
40
20

The Magician's Apprentice    6.5
The Witch's Familiar    5.6    6.7
Under the Lake    6.1
Before the Flood    6.6    5.6
The Girl Who Died    6.1
The Woman Who Lived    5.8
The Zygon Invasion    6
The Zygon Inversion    5.6
Sleep No More
Face the Raven    6.1
Heaven Sent    6.2
Hell Bent    6.2
The Husbands of River Song    7.7

193

# THE BEST OF TIMES,
## THE WORST OF TIMES...

When to show Doctor Who
to get the most people watching.

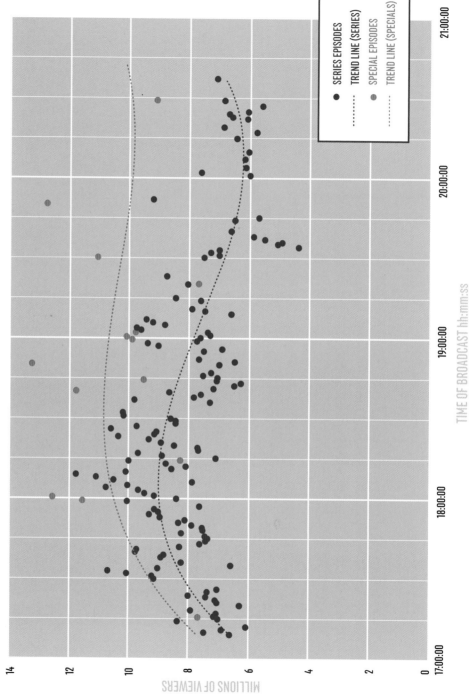

**AVERAGE VIEWING FIGURES BY TIME OF BROADCAST**
(SPECIALS EXCLUDED)

MILLIONS OF VIEWERS

TIME OF BROADCAST hh:mm:ss

Legend:
- SERIES EPISODES
- TREND LINE (SERIES)
- SPECIAL EPISODES
- TREND LINE (SPECIALS)

# AVERAGE VIEWING FIGURES BY MONTH
## (SERIES vs SPECIALS)

**MILLIONS OF VIEWERS**

14
12
10
8
6
4
2
0

JAN  FEB  MAR  APR  MAY  JUN  JUL  AUG  SEP  OCT  NOV  DEC

MONTH OF BROADCAST

- ● SERIES EPISODES
- ⋯ TREND LINE (SERIES)
- ● SPECIAL EPISODES
- ⋯ TREND LINE (SPECIALS)

OPTIMUM TIME (SERIES EPISODE) = **18:15**    OPTIMUM TIME (SPECIAL EPISODE) = **18:30**    OPTIMUM MONTH (SERIES/ SPECIAL) = **DEC or JAN**

* Special "event" episodes – The Five Doctors (1983), Doctor Who (1996), the Christmas specials since 2005, Planet of the Dead (2009),
The Waters of Mars (2009), The End of Time part 2 (2010) and The Day of the Doctor (2013) – skew the figures, so we've shown them separately.

195

# MISSING EPISODES

Of the 826 episodes of Doctor Who to date, 97 are missing from the BBC archive, all from the First and Second Doctors' eras.

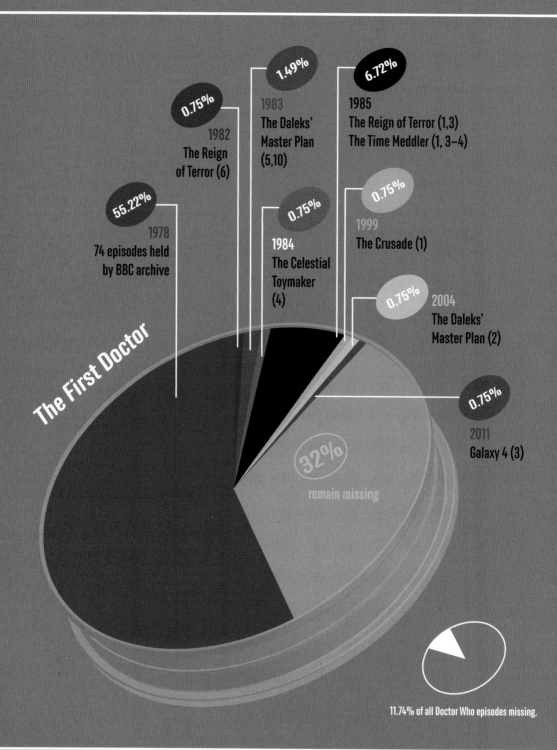

**0.75%**
1982
The Reign
of Terror (6)

**1.49%**
1983
The Daleks'
Master Plan
(5,10)

**6.72%**
1985
The Reign of Terror (1,3)
The Time Meddler (1, 3–4)

**55.22%**
1978
74 episodes held
by BBC archive

**0.75%**
1984
The Celestial
Toymaker
(4)

**0.75%**
1999
The Crusade (1)

**0.75%**
2004
The Daleks'
Master Plan (2)

**0.75%**
2011
Galaxy 4 (3)

The First Doctor

**32%**
remain missing

11.74% of all Doctor Who episodes missing.

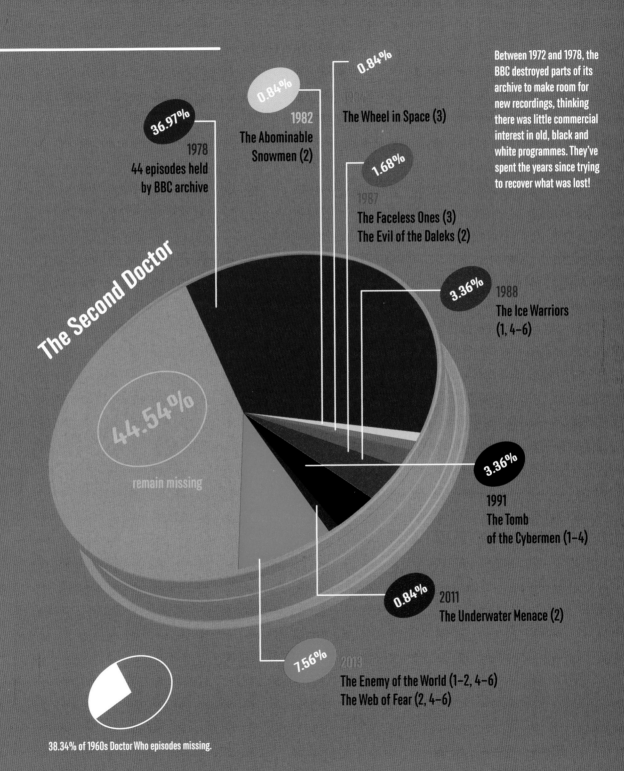

**The Second Doctor**

36.97%

1978
44 episodes held
by BBC archive

0.84%
1982
The Abominable
Snowmen (2)

0.84%
1974
The Wheel in Space (3)

1.68%
1987
The Faceless Ones (3)
The Evil of the Daleks (2)

3.36%
1988
The Ice Warriors
(1, 4–6)

Between 1972 and 1978, the
BBC destroyed parts of its
archive to make room for
new recordings, thinking
there was little commercial
interest in old, black and
white programmes. They've
spent the years since trying
to recover what was lost!

44.54%
remain missing

3.36%
1991
The Tomb
of the Cybermen (1–4)

0.84%
2011
The Underwater Menace (2)

7.56%
2013
The Enemy of the World (1–2, 4–6)
The Web of Fear (2, 4–6)

38.34% of 1960s Doctor Who episodes missing.

**1** TICKET TO RIDE
THE BEATLES
THE CHASE

**1** PAPERBACK WRITER
THE BEATLES
THE EVIL OF THE DALEKS

**1** A WHITER SHADE OF PALE
PROCOL HARUM
REVELATION OF THE DALEKS

**1** TAINTED LOVE
SOFT CELL
THE END OF THE WORLD

**1** TOXIC
BRITNEY SPEARS
THE END OF THE WORLD

**1** NEVER GONNA GIVE YOU UP
RICK ASTLEY
FATHER'S DAY

**1** MERRY CHRISTMAS EVERYBODY
SLADE
THE CHRISTMAS INVASION, THE RUNAWAY BRIDE AND LAST CHRISTMAS

**1** HIT ME WITH YOUR RHYTHM STICK
IAN DURY AND THE BLOCKHEADS
TOOTH AND CLAW

**1** THE LION SLEEPS TONIGHT
TIGHT FIT
RISE OF THE CYBERMEN

**1** YOU DON'T HAVE TO SAY YOU LOVE ME
DUSTY SPRINGFIELD
THE REBEL FLESH

**1** FEEL THE LOVE
RUDIMENTAL FEATURING JOHN NEWMAN
ASYLUM OF THE DALEKS

**1** TITANIUM
DAVID GUETTA FEATURING SIA
THE POWER OF THREE

**1** GHOST TOWN
THE SPECIALS
THE RINGS OF AKHATEN

**2** OH WELL
FLEETWOOD MAC
SPEARHEAD FROM SPACE

**2** LOLLIPOP
THE MUDLARKS
REMEMBRANCE OF THE DALEKS

**2** VIENNA
ULTRAVOX
COLD WAR

**2** ROLLING IN THE DEEP
ADELE
THE IMPOSSIBLE ASTRONAUT

**2** THE BIRDIE SONG
THE TWEETS
THE POWER OF THREE

**3** DON'T BRING ME DOWN
ELO
LOVE & MONSTERS

**3** COULD IT BE MAGIC
TAKE THAT
PARTNERS IN CRIME

**3** VOODOO CHILD
ROUGH TRADERS
THE SOUND OF DRUMS

NEVER CAN SAY GOODBYE
THE COMMUNARDS
**4** FATHER'S DAY

**4** DANIEL
ELTON JOHN
LOVE & MONSTERS

SUPERMASSIVE BLACK HOLE
**4** MUSE
THE REBEL FLESH

Highest UK chart position.

# HIT PARADE

| | | | |
|---|---|---|---|
| YOU GIVE ME SOMETHING<br>**JAMES MORRISON**<br>**5** THE BIG BANG | **15** ENGLISHMAN<br>IN NEW YORK<br>**STING**<br>THE ANGELS<br>TAKE MANHATTAN | ALBUM TRACKS<br>↓ | TWENTY-FOUR<br>HOURS FROM TULSA<br>**DUSTY SPRINGFIELD**<br>PARTNERS IN CRIME |
| HUNGRY LIKE A WOLF<br>**DURAN DURAN**<br>**5** COLD WAR | **15** FIRE WOMAN<br>**THE CULT**<br>JOURNEY TO THE<br>CENTRE OF THE TARDIS | THE DEVIL'S TRIANGLE<br>**KING CRIMSON**<br>THE MIND OF EVIL | CHANCE<br>**ATHLETE**<br>VINCENT AND<br>THE DOCTOR |
| MR BLUE SKY<br>**6** ELO<br>LOVE & MONSTERS | **18** TURN TO STONE<br>**ELO**<br>LOVE & MONSTERS | TANK<br>**EMERSON, LAKE &<br>PALMER**<br>COLONY IN SPACE | FIRE<br>**JIMI HENDRIX**<br>REVELATION<br>OF THE DALEKS |
| RAFFAELLA CARRA<br>DO IT, DO IT AGAIN<br>**9** MIDNIGHT | **21** DON'T MUG YOURSELF<br>**THE STREETS**<br>FATHER'S DAY | DO YOU WANT TO<br>KNOW A SECRET?<br>**THE BEATLES**<br>REMEMBRANCE<br>OF THE DALEKS | UNKNOWN<br>↓ |
| STARMAN<br>**10** DAVID BOWIE<br>ALIENS OF LONDON | **43** CRAZY LITTLE THING<br>CALLED LOVE<br>**QUEEN**<br>THE BIG BANG | A TASTE OF HONEY<br>**THE BEATLES**<br>REMEMBRANCE<br>OF THE DALEKS | NOBODY KNOWS THE<br>TROUBLE I'VE SEEN<br>**THE SEEKERS**<br>THE EVIL OF THE DALEKS |
| LOVE WILL<br>TEAR US APART<br>**13** JOY DIVISION<br>SCHOOL REUNION | **64** I CAN'T DECIDE<br>**SCISSOR SISTERS**<br>LAST OF THE TIME LORDS | REGRESA A MI<br>**IL DIVO**<br>LOVE & MONSTERS | |

"That's what you are:
a big old punk, with a
bit of rockabilly thrown in."
→ Rose Tyler,
Tooth and Claw (2006)

# TIME OF **THE DOCTOR**

At what hour the 826 episodes of Doctor Who have been shown.

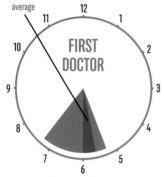

**EARLIEST START TIME** – THE ESCAPE (AND OTHERS) – 17:15:00
**LATEST FINISH TIME** – THE WAR MACHINES EPISODE 2 – 19:20:00

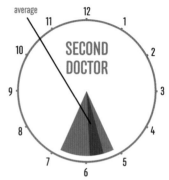

**EARLIEST START TIME** – THE ICE WARRIORS EPISODE 1 – 17:10:00
**LATEST FINISH TIME** – THE EVIL OF THE DALEKS EPISODE 7 – 18:49:33

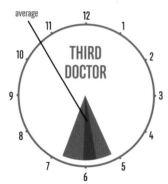

**EARLIEST START TIME** – THE TIME WARRIOR PART 1 – 17:10:00
**LATEST FINISH TIME** – THE DÆMONS EPISODE 1 – 18:42:05

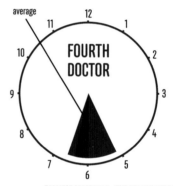

**EARLIEST START TIME** – THE KEEPER OF TRAKEN PART 1 (AND OTHERS) – 17:09:00
**LATEST FINISH TIME** – THE TALONS OF WENG-CHIANG PART 2 – 19:01:26

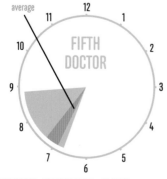

**EARLIEST START TIME** – FRONTIOS PART 3 – 18:40:00
**LATEST FINISH TIME** – THE FIVE DOCTORS – 20:50:23

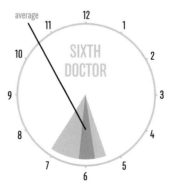

**EARLIEST START TIME** – THE TRIAL OF A TIME LORD PART 13 – 17:20:00
**LATEST FINISH TIME** – THE TWIN DILEMMA PART 4 – 19:08:04

"I just love clocks: atomic clocks, wall quartz clocks, grandfather clocks ... Cuckoo clocks."
→ The Fourth Doctor, The Sontaran Experiment (1975)

EARLIEST START TIME – TIME AND THE RANI PART 2 (AND OTHERS) – 19:34:00
LATEST FINISH TIME – THE GREATEST SHOW IN THE GALAXY PART 3 – 20:04:30

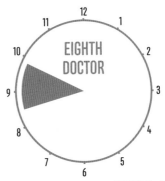

EARLIEST START TIME – DOCTOR WHO – 20:29:00
LATEST FINISH TIME – DOCTOR WHO – 21:53:39

EARLIEST START TIME – THE EMPTY CHILD – 18:29:00
LATEST FINISH TIME – THE UNQUIET DEAD – 19:44:48

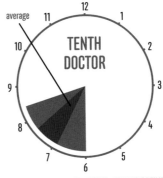

EARLIEST START TIME – THE END OF TIME PART 1 – 17:59:00
LATEST FINISH TIME – GRIDLOCK – 20:24:58

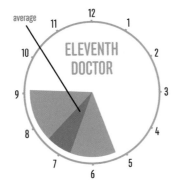

EARLIEST START TIME – THE SNOWMEN – 17:14:00
LATEST FINISH TIME – THE DAY OF THE DOCTOR – 21:06:31

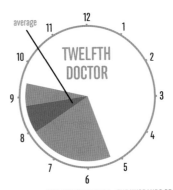

EARLIEST START TIME – THE HUSBANDS OF RIVER SONG – 17:15:00
LATEST FINISH TIME – MUMMY ON THE ORIENT EXPRESS – 21:23:31

# Jobs
## FOR Women

Ratios of women in key roles
in the production of Doctor Who.

0.24% of episodes
have music
composed by women

2.72% of episodes
written by women

7.51%
of episodes
directed by women

8.96%
of episodes
designed by women

14.53%
of episodes produced by women
(not including executive producers
or series producers)

**99.76%**
of episodes have make-up/
make-up design by women

**62.69%**
of episodes have
costumes designed
by women

826 episodes of Doctor Who

# THE DAY OF THE DOCTOR

*Guinness World Record holder for the largest ever simulcast of a TV drama (23 November 2013, 19:50 GMT)*

> "I understand you're rather fond of this world."
> — Elizabeth I to the Doctor, The Day of the Doctor (2013)

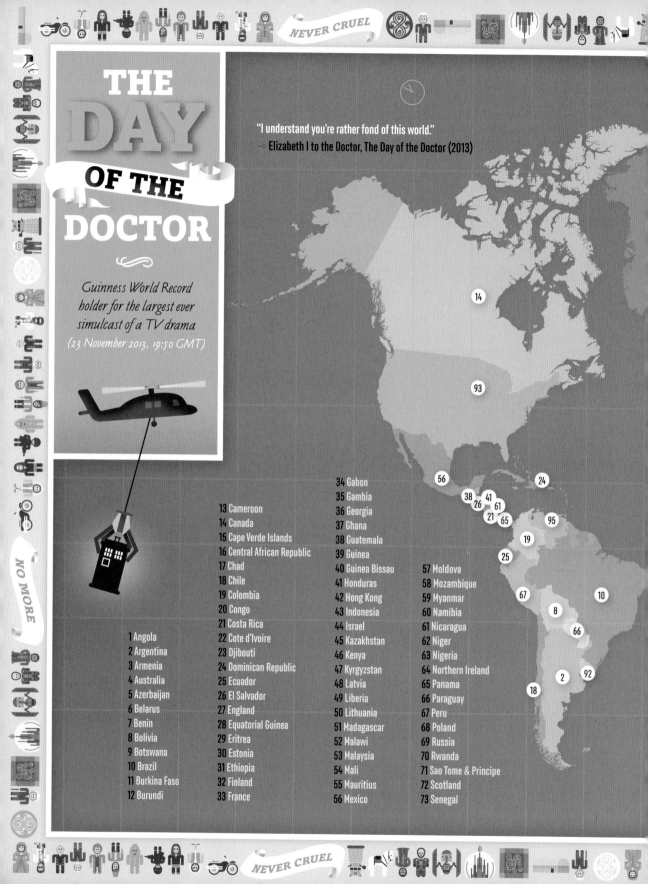

1 Angola
2 Argentina
3 Armenia
4 Australia
5 Azerbaijan
6 Belarus
7 Benin
8 Bolivia
9 Botswana
10 Brazil
11 Burkina Faso
12 Burundi
13 Cameroon
14 Canada
15 Cape Verde Islands
16 Central African Republic
17 Chad
18 Chile
19 Colombia
20 Congo
21 Costa Rica
22 Cote d'Ivoire
23 Djibouti
24 Dominican Republic
25 Ecuador
26 El Salvador
27 England
28 Equatorial Guinea
29 Eritrea
30 Estonia
31 Ethiopia
32 Finland
33 France
34 Gabon
35 Gambia
36 Georgia
37 Ghana
38 Guatemala
39 Guinea
40 Guinea Bissau
41 Honduras
42 Hong Kong
43 Indonesia
44 Israel
45 Kazakhstan
46 Kenya
47 Kyrgyzstan
48 Latvia
49 Liberia
50 Lithuania
51 Madagascar
52 Malawi
53 Malaysia
54 Mali
55 Mauritius
56 Mexico
57 Moldova
58 Mozambique
59 Myanmar
60 Namibia
61 Nicaragua
62 Niger
63 Nigeria
64 Northern Ireland
65 Panama
66 Paraguay
67 Peru
68 Poland
69 Russia
70 Rwanda
71 Sao Tome & Principe
72 Scotland
73 Senegal

NEVER CRUEL

NO MORE

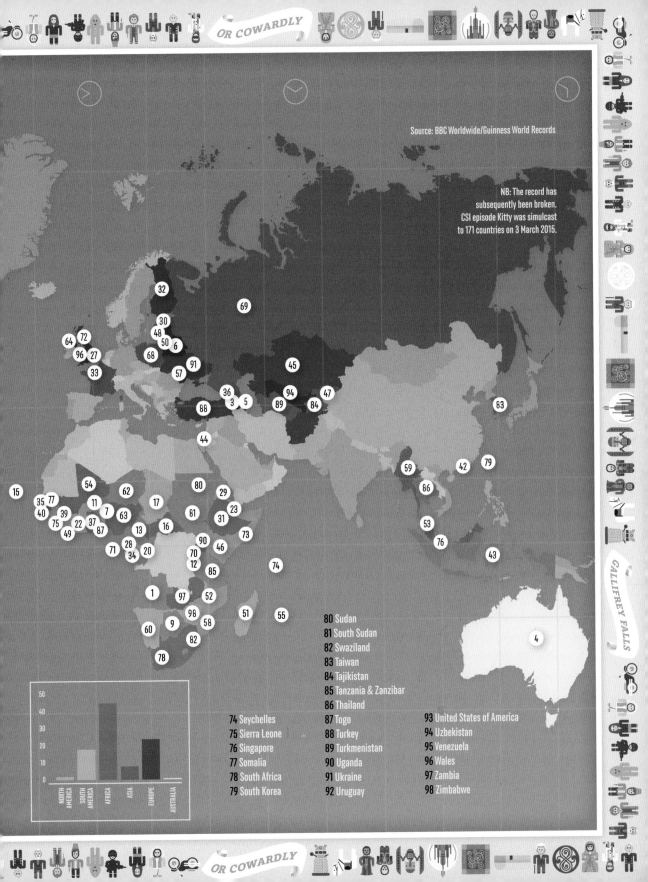

OR COWARDLY

Source: BBC Worldwide/Guinness World Records

NB: The record has subsequently been broken. CSI episode Kitty was simulcast to 171 countries on 3 March 2015.

GALLIFREY FALLS

80 Sudan
81 South Sudan
82 Swaziland
83 Taiwan
84 Tajikistan
85 Tanzania & Zanzibar
86 Thailand

74 Seychelles
75 Sierra Leone
76 Singapore
77 Somalia
78 South Africa
79 South Korea

87 Togo
88 Turkey
89 Turkmenistan
90 Uganda
91 Ukraine
92 Uruguay

93 United States of America
94 Uzbekistan
95 Venezuela
96 Wales
97 Zambia
98 Zimbabwe

50
40
30
20
10
0

NORTH AMERICA | SOUTH AMERICA | AFRICA | ASIA | EUROPE | AUSTRALIA

OR COWARDLY

# And Finally...

16 DAYS, 8 HOURS,
13 MINUTES, 51 SECONDS
The total duration
of all 826 episodes
of Doctor Who

**27%**

MORE THAN ONE QUARTER –
222/826 – OF DOCTOR WHO
EPISODES WERE SHOWN IN
ITS FIRST FIVE YEARS.

AVERAGE
LENGTH OF
THE 826
EPISODES:
28 MINUTES,
28 SECONDS

LONGEST
EPISODE –
THE FIVE
DOCTORS
(25 NOVEMBER
1983) –
1 HOUR,
30 MINUTES
AND 23 SECONDS

SHORTEST EPISODE –
THE MIND ROBBER
EPISODE 5
(12 OCTOBER 1968) –
18 MINUTES

**29 MINUTES,
30 SECONDS**

THE TRIAL OF A TIME LORD
PART 14 (6 DECEMBER 1986) –

**EPISODE CLOSEST
TO AVERAGE LENGTH**

**7.96 MILLION**

**AVERAGE VIEWERS OF
THE 826 EPISODES**

EPISODE
WITH
HIGHEST
VIEWING
FIGURES:
CITY OF DEATH
PART 4
(20 OCTOBER
1979) –
16.1 MILLION

EPISODE WITH
LOWEST VIEWING
FIGURES:
BATTLEFIELD PART 1
(6 SEPTEMBER 1989) –
3.1 MILLION

# FOURTH DOCTOR
(Robot part 1 to Logopolis part 4)
## 2 DAYS, 21 HOURS,
### 6 MINUTES, 54 SECONDS

# FIRST DOCTOR
(An Unearthly Child to
The Tenth Planet episode 4)
## 2 DAYS, 6 HOURS,
### 24 MINUTES, 37 SECONDS

# THIRD DOCTOR
(Spearhead from Space episode 1 to
Planet of the Spiders episode 6)
## 2 DAYS, 3 HOURS,
### 53 MINUTES, 28 SECONDS

# SECOND DOCTOR
(The Power of the Daleks episode 1 to
The War Games episode 10)
## 1 DAY, 23 HOURS,
### 19 MINUTES, 12 SECONDS

# TENTH DOCTOR
(The Christmas Invasion to
The End of Time part 2)
## 1 DAY, 14 HOURS,
### 11 MINUTES, 8 SECONDS

# ELEVENTH DOCTOR
(The Eleventh Hour to
The Time of the Doctor)
## 1 DAY, 10 HOURS,
### 52 MINUTES

# FIFTH DOCTOR
(Castrovalva part 1 to
The Caves of Androzani part 4)
## 1 DAY, 9 HOURS,
### 59 MINUTES, 53 SECONDS

# TWELFTH DOCTOR
– so far (Deep Breath
to The Husbands of River Song)
## 21 HOURS,
### 15 MINUTES, 45 SECONDS

# SIXTH DOCTOR
(The Twin Dilemma part 1
to The Trial of a Time Lord part 14)
## 17 HOURS,
### 9 MINUTES, 56 SECONDS

# SEVENTH DOCTOR
(Time and the Rani part 1
to Survival part 3)
## 17 HOURS,
### 5 MINUTES, 31 SECONDS

# NINTH DOCTOR
(Rose to
The Parting of the Ways)
## 9 HOURS,
### 30 MINUTES, 38 SECONDS

# EIGHTH DOCTOR
(Doctor Who)
## 1 HOUR,
### 24 MINUTES, 39 SECONDS

# THE MAKING OF WHOGRAPHICA

OUR YEAR CREATING THE BOOK

## TIMELINE OF WHOGRAPHICA

**20 OCTOBER 2014**
Steve contacts BBC Books with a proposal for a book of Doctor Who infographics.

**23 FEBRUARY 2015**
Steve emails Justin Richards at BBC Books with a sample chapter.

**20 APRIL 2015**
Justin asks Simon if he would be interested in collaborating with Steve and Ben.

**21 APRIL 2015**
Simon sends a chapter breakdown based on The Scientific Secrets of Doctor Who.

**JULY/ AUGUST 2015**
Ben researches how to create many different types of infographics, and designs two sample spreads.

**8 JULY 2015**
Ben meets Albert DePetrillo, Senior Editorial Director at BBC Books, to discuss ideas for the project.

**JUNE 2015**
Steve and Simon revise the pitch.

**26 MAY 2015**
A formal pitch is sent to BBC Books. The book is called The Secret Knowledge of Doctor Who at this stage.

**13 MAY 2015**
Simon begins a huge spreadsheet of Doctor Who episodes, listing dates of broadcast, start times, length of episodes etc.

**26 AUGUST 2015**
Ben meets Albert at the Edinburgh International Book Festival to discuss further ideas. The title of the book is now Whographica.

**8 SEPTEMBER 2015**
Charlotte Macdonald is appointed as project manager for the project.

**9 SEPTEMBER 2015**
Steve, Simon and Ben are in touch about Whographica for the first time.

**19 SEPTEMBER 2015**
Steve and Simon show Justin progress on the book at the Big Finish Day convention.

**27 NOVEMBER 2015**
A flatplan is sent to the BBC for approval.

**JANUARY 2016**
Toby Clarke is commissioned to design the cover of the book.

**22 SEPTEMBER 2016**
Publication date!

**5 JANUARY 2016**
A synopsis for the book appears on the BBC Books website.

**7 APRIL 2016**
Book jacket and spreads to production.

**5 JANUARY 2016**
A revised flatplan from Simon.

**24 MARCH 2016**
Amendments required by the Doctor Who production office and proofreaders are completed.

**1 FEBRUARY 2016**
The first batch of completed designs is submitted to Charlotte. Two-thirds of Whographica is now complete.

**8 MARCH 2016**
The final spread, The Day of the Doctor simulcast map, is completed. Steve Tribe checks the last batch of finished designs.

**10 FEBRUARY 2016**
The first batch of notes is received from Steve Tribe, via Grace Paul at BBC Books.

**26 FEBRUARY 2016**
Simon is a guest on Doctor Who: The Fan Show, where the book gets its first promotional plug!

**15 FEBRUARY 2016**
All fonts are approved by BBC Books and licenses for them are acquired.

END OF PROJECT

STRESS LEVELS

WHOGRAPHICA

BEGINNING OF PROJECT

Six months of long days, late nights and few days off, writing and designing Whographica!

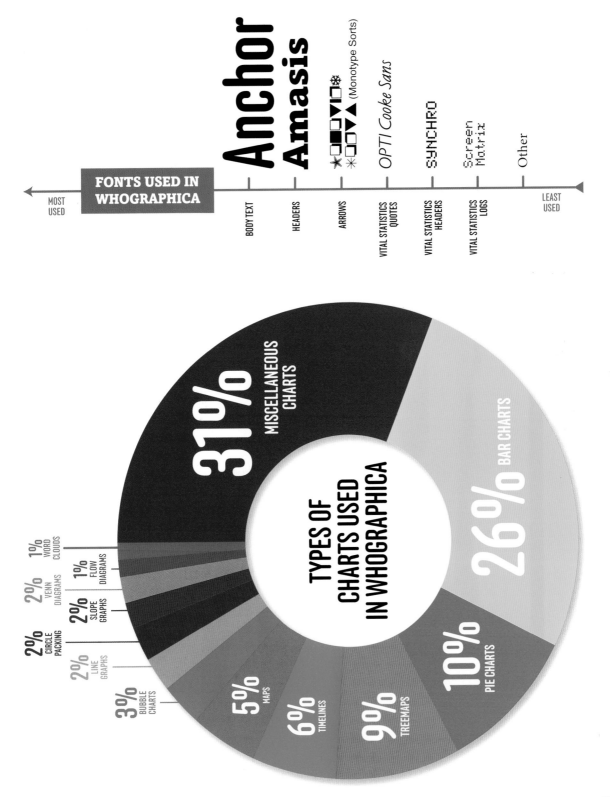

FONTS USED IN WHOGRAPHICA

MOST USED

LEAST USED

Anchor — BODY TEXT
Amasis — HEADERS
◼◻▮▼◺◻ (Monotype Sorts) — ARROWS
OPTI Cooke Sans — VITAL STATISTICS QUOTES
SYNCHRO — VITAL STATISTICS HEADERS
Screen Matrix — VITAL STATISTICS LOGS
Other — 

TYPES OF CHARTS USED IN WHOGRAPHICA

31% MISCELLANEOUS CHARTS
26% BAR CHARTS
10% PIE CHARTS
9% TREEMAPS
6% TIMELINES
5% MAPS
3% BUBBLE CHARTS
2% LINE GRAPHS
2% CIRCLE PACKING
2% SLOPE GRAPHS
2% VENN DIAGRAMS
1% FLOW DIAGRAMS
1% WORD CLOUDS

# You've Got Mail

- STEVE
- SIMON
- BEN

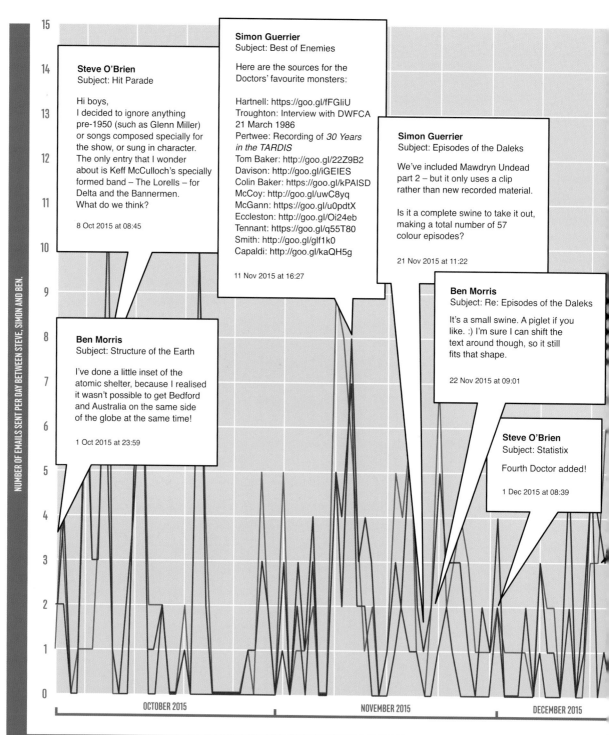

**Steve O'Brien**
Subject: Hit Parade

Hi boys,
I decided to ignore anything pre-1950 (such as Glenn Miller) or songs composed specially for the show, or sung in character. The only entry that I wonder about is Keff McCulloch's specially formed band – The Lorells – for Delta and the Bannermen. What do we think?

8 Oct 2015 at 08:45

**Simon Guerrier**
Subject: Best of Enemies

Here are the sources for the Doctors' favourite monsters:

Hartnell: https://goo.gl/fFGliU
Troughton: Interview with DWFCA 21 March 1986
Pertwee: Recording of *30 Years in the TARDIS*
Tom Baker: http://goo.gl/22Z9B2
Davison: http://goo.gl/iGEIES
Colin Baker: https://goo.gl/kPAISD
McCoy: http://goo.gl/uwC8yq
McGann: https://goo.gl/u0pdtX
Eccleston: http://goo.gl/Oi24eb
Tennant: https://goo.gl/q55T80
Smith: http://goo.gl/glf1k0
Capaldi: http://goo.gl/kaQH5g

11 Nov 2015 at 16:27

**Simon Guerrier**
Subject: Episodes of the Daleks

We've included Mawdryn Undead part 2 – but it only uses a clip rather than new recorded material.

Is it a complete swine to take it out, making a total number of 57 colour episodes?

21 Nov 2015 at 11:22

**Ben Morris**
Subject: Structure of the Earth

I've done a little inset of the atomic shelter, because I realised it wasn't possible to get Bedford and Australia on the same side of the globe at the same time!

1 Oct 2015 at 23:59

**Ben Morris**
Subject: Re: Episodes of the Daleks

It's a small swine. A piglet if you like. :) I'm sure I can shift the text around though, so it still fits that shape.

22 Nov 2015 at 09:01

**Steve O'Brien**
Subject: Statistix

Fourth Doctor added!

1 Dec 2015 at 08:39

NUMBER OF EMAILS SENT PER DAY BETWEEN STEVE, SIMON AND BEN.

OCTOBER 2015    NOVEMBER 2015    DECEMBER 2015

DURATION OF THE COLLABORATION BETWEEN STEVE, SIMON AND BEN ON WHOGRAPHICA

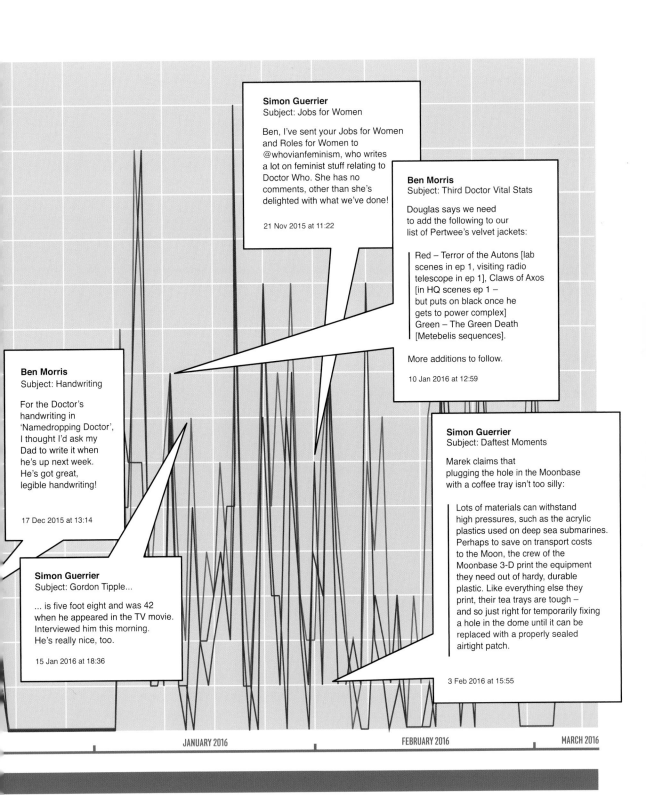

**Simon Guerrier**
Subject: Jobs for Women

Ben, I've sent your Jobs for Women and Roles for Women to @whovianfeminism, who writes a lot on feminist stuff relating to Doctor Who. She has no comments, other than she's delighted with what we've done!

21 Nov 2015 at 11:22

**Ben Morris**
Subject: Third Doctor Vital Stats

Douglas says we need to add the following to our list of Pertwee's velvet jackets:

Red – Terror of the Autons [lab scenes in ep 1, visiting radio telescope in ep 1], Claws of Axos [in HQ scenes ep 1 – but puts on black once he gets to power complex]
Green – The Green Death [Metebelis sequences].

More additions to follow.

10 Jan 2016 at 12:59

**Ben Morris**
Subject: Handwriting

For the Doctor's handwriting in 'Namedropping Doctor', I thought I'd ask my Dad to write it when he's up next week. He's got great, legible handwriting!

17 Dec 2015 at 13:14

**Simon Guerrier**
Subject: Daftest Moments

Marek claims that plugging the hole in the Moonbase with a coffee tray isn't too silly:

Lots of materials can withstand high pressures, such as the acrylic plastics used on deep sea submarines. Perhaps to save on transport costs to the Moon, the crew of the Moonbase 3-D print the equipment they need out of hardy, durable plastic. Like everything else they print, their tea trays are tough – and so just right for temporarily fixing a hole in the dome until it can be replaced with a properly sealed airtight patch.

3 Feb 2016 at 15:55

**Simon Guerrier**
Subject: Gordon Tipple...

... is five foot eight and was 42 when he appeared in the TV movie. Interviewed him this morning. He's really nice, too.

15 Jan 2016 at 18:36

JANUARY 2016                    FEBRUARY 2016                    MARCH 2016

# ACKNOWLEDGEMENTS

Grateful thanks to the clever people who checked our calculations and pointed out things we got wrong. (If we're still wrong, that's our fault, not theirs.)

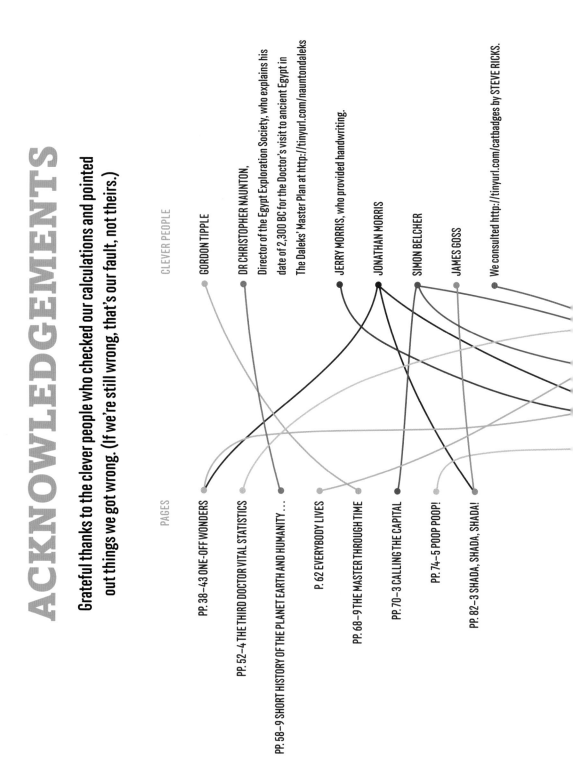

PAGES

CLEVER PEOPLE

PP. 38–43 ONE-OFF WONDERS

PP. 52–4 THE THIRD DOCTOR VITAL STATISTICS

PP. 58–9 SHORT HISTORY OF THE PLANET EARTH AND HUMANITY . . .

P. 62 EVERYBODY LIVES

PP. 68–9 THE MASTER THROUGH TIME

PP. 70–3 CALLING THE CAPITAL

PP. 74–5 POOP POOP!

PP. 82–3 SHADA, SHADA, SHADA!

GORDON TIPPLE

DR CHRISTOPHER NAUNTON,
Director of the Egypt Exploration Society, who explains his date of 2,300 BC for the Doctor's visit to ancient Egypt in The Daleks' Master Plan at http://tinyurl.com/nauntondaleks

JERRY MORRIS, who provided handwriting.

JONATHAN MORRIS

SIMON BELCHER

JAMES GOSS

We consulted http://tinyurl.com/catbadges by STEVE RICKS.

ALYSSA FRANKE @WhovianFeminism

DR MAREK KUKULA,
Public Astronomer at the Royal Observatory Greenwich

The Target Book (2008) by DAVID J HOWE.

TOBY HADOKE

DR DOUGLAS MCNAUGHTON

At BBC Worldwide: CHRIS ALLEN,
JAMES DUDLEY and JULIA NOCCIOLINO

CHRISTEL DEE and LUKE SPILLANE from the Doctor Who Fan Show

ANDREW LEDGER

Where dates are not given on screen, we referred to
the chronology in AHistory: An Unauthorized History
of the Doctor Who Universe (2006) by LANCE PARKIN,
and consulted its publisher, LARS PEARSON.

@ BBC Books: ALBERT DEPETRILLO, GRACE PAUL, TESSA HENDERSON,
CHARLOTTE MACDONALD, JUSTIN RICHARDS, STEVE TRIBE, PAUL SIMPSON

P. 85 DOCTOR WHO'S DAFTEST MOMENTS

PP. 90–1 ON TARGET

PP. 104–5 SARAH JANE'S ADVENTURES

PP. 108–9 ROLES FOR WOMEN

PP. 110–1 THE NAME-DROPPING DOCTOR

PP. 118–9 THE SIXTH DOCTOR VITAL STATISTICS

PP. 168–175 EVERYONE WE'VE EVER SEEN INSIDE THE TARDIS

PP. 184–5 PLANETARY ARRIVALS

PP. 188–9 WHERE IS GALLIFREY?

PP. 202–3 JOBS FOR WOMEN

PP.204–5 THE DAY OF THE DOCTOR

# INDEX

Doctor Who: Whographica
Artwork © Ben Morris 2016
Text © Simon Guerrier and Steve O'Brien 2016

HarperCollins books may be purchased for educational,
business, or sales promotional use. For information please e-mail
the Special Markets Department at SPsales@harpercollins.com.

First published in 2016 by
Harper Design,
*An Imprint of* HarperCollins*Publishers*
195 Broadway
New York, NY 10007
Tel: (212) 207-7000
Fax: 855-746-6023
harperdesign@harpercollins.com
www.hc.com

This edition distributed throughout the world by:
HarperCollins*Publishers*
195 Broadway
New York, NY 10007

Library of Congress Control Number: 2016935760

ISBN: 978-0-06-247022-5

First Printing, 2016

EDITORIAL DIRECTOR: Albert DePetrillo
EDITOR: Charlotte Macdonald
SERIES CONSULTANT: Justin Richards
COPYEDITOR: Steve Tribe
PRODUCTION: Phil Spencer
DESIGNER: Jim Smith

Printed and bound in China by Toppan Leefung